KA 0338260 5

KV-051-304

WITHDRAWN FROM
WINCHESTER

Buying Knowledge

To Joyce and Mark, for their patience and forbearance, without which this book would not have been possible.

Buying Knowledge

Effective Acquisition of External Knowledge

PETER SAMMONS

GOWER

Published by
Gower Publishing Limited
Gower House
Croft Road
Aldershot
Hants GU11 3HR
England

Gower Publishing Company
Suite 420
101 Cherry Street
Burlington,
VT 05401-4405
USA

British Library Cataloguing in Publication Data
Sammons, Peter A.
Buying knowledge : effective acquisition of external knowledge
1. Knowledge management 2. Intellectual captial
3. Intellectual property 4. Information resources - Economic aspects 5. Communication of technical information - Economic aspects
I. Title
658.4'038

ISBN 0 566 08635 2

Library of Congress Cataloging-in-Publication Data
Sammons, Peter A.
Buying knowledge : effective acquisition of external knowledge / by Peter Sammons.
p. cm.
Includes index.
ISBN: 0-566-08635-2
1. Intellectual capital. 2. Intellectual property. 3. Information resources--Economic aspects. 4.
Communication of technical information--Economic aspects. I. Title.
HD53.S25 2004
658.4'038--dc22

2004020941

Typeset by IML Typographers, Birkenhead, Merseyside.
Printed and bound in Great Britain by MPG Books Ltd., Bodmin.

Contents

List of Figures

List of Tables

Introduction

WHY A BOOK ABOUT BUYING KNOWLEDGE?

The twentieth century may, in retrospect, be viewed as *the century of exponential growth of knowledge.* It was the century when mankind conquered the air, space, the deep ocean; when the human genome was unravelled and entire new industries were developed – the film industry, electronics, radio, telecommunications. It was the century in which medicine developed from the pre-penicillin era, when amputation was often the only solution for surgical complications, to the era of keyhole surgery and magnetic resonance imaging. Education for the masses – in the Western world at any rate (and for many of these, a good education up to tertiary levels) became a reality. Knowledge of history, geography and the sciences became available to a majority of people in the industrially developed world, and the electronic computer became ubiquitous.

From the second half of the twentieth century science and technology advanced with ever greater momentum. Towards the end of the century, in business, government and all large organizations, those who led these organizations – managers by whatever title they were known – were required to be familiar with an ever-increasing and often incomprehensible array of information. Social advance brought ever greater regulatory control. The digital age brought with it vast quantities of data in a bewildering array of guises. Every element of organizational management became 'professionalized'.

No longer could managers – or indeed employees – work and think in professional silos. Instead, cross-functional working became the norm. Employee flexibility, no less than organizational flexibility, became essential to survival, let alone success. Organizations increasingly became required to manage a vast array of data inputs in order to develop usable information. Power and responsibility were delegated ever lower in organizations, as senior managers found they could no longer micro-manage every aspect of their organization. Decisions increasingly became based on information and knowledge – rather than hunch and precedent. Managers were required to manage more and more, in ever decreasing timescales. Resources were often reduced as organizations 'right-sized'.

For many managers and employees, there came an added responsibility to spend (or invest) their organization's revenue to obtain usable knowledge. Organizations were subject to increasing need to acquire and utilize new knowledge.

For some within the professions, experts in their own right, identifying knowledge gaps and diligently searching out knowledge in usable, dependable forms, became a normal part of their working life. But stories of mismanagement and failed projects suggested that organizations were all too often unable to identify knowledge needs/gaps, or to manage that knowledge effectively when it became available. Was this a failure of basic business nous? Of management? Was this a failure of the (relatively new) profession of 'purchasing'?

In the twenty-first century, it seems likely that knowledge will continue to grow steadily. Organizations will need to remain abreast of technical, social and legal developments. To do this, some knowledge will be developed within the organization, whilst the rest will be sourced externally – and from this emerges the concept of *buying knowledge*. It is not simply another aspect of purchasing, or something that can be left to a buyer and a good old-fashioned competitive tender. The process of buying knowledge is a team-based activity as organizations determine what new knowledge is needed and then how to access, acquire, transfer and finally internalize that knowledge.

It seems likely that most managers at some point in their career will assume responsibility for a project to *buy* new knowledge – or less commonly, to buy old knowledge in a new form – and projects using external resources will increasingly involve some element of knowledge creation, capture and internal transfer. The purpose of this book is to set out some basic building blocks which organizations can use to recognize when a project is primarily about knowledge creation, and accordingly plan projects in such a way that the knowledge transfer element is given due importance.

This is the author's second business book. The first, also published by Gower, was *The Outsourcing R&D Toolkit*, published in August 2000 ('R&D' refers to scientific or technological research and development). That book emerged from the author's work in the field of commercial contracts in R&D-intensive industries. It focused on the processes, procedures and pitfalls organizations encounter when buying technological R&D services and contained a 'toolbox' of commercial materials that organizations could consider using to support their external R&D sourcing.

This book tackles a related subject. Buying knowledge-based services, whether by 'outsourcing' or other commercial mechanisms, is a more generic activity than buying technological R&D. The author presently works for a global financial services firm, involved in the acquisition of professional and other services. These services often involve the creation of new knowledge, which has been created at the expense of the client organization, and to the requirements of the client organization. Sometimes the knowledge is unique, and would not have been developed except at the instigation of the client. Why is it, then, that the client often concludes that it has not realized value for money in the receipt of the services? Or that although the services themselves are satisfactory, the client never really acquires true *ownership* of them? And why is it that all too often the organization finds itself having to buy in the same or similar services in the future? It is as if the organization has failed to learn from its past experience.

WHY MANAGERS GET IT WRONG

This book aims to help managers avoid traps in spending their organization's money with external suppliers of knowledge. Managers of course need to take responsibility for the execution of projects and (yes, yet again) to keep abreast of information about why projects go wrong and managers fail. In a cautionary article in *Harvard Business Review* titled 'The harder they fall'[1] Roderick Kramer builds the case that a disproportionate number of leaders in politics, business, the media and even religion, display remarkable adeptness and shrewdness whilst working their way up the career ladder, only to fall prey to what the author calls 'bouts of folly' once they achieve very senior positions. Roderick Kramer believes there is something in the process of becoming leader that changes behaviours in individuals in quite profound ways. The systems through which leaders are identified, groomed and exposed to career development opportunities, require managers to sacrifice the attitudes essential to survival once they have reached the top. Society at large has come to the view that risk-taking and rule bending are markers of good leadership. As a result, senior managers lack the caution and modesty needed to cope with the trappings of power, believing that normal rules do not apply to them and that they are entitled to seize all the rewards that power makes available to them.

A fascinating article, and one well worth reading by anyone determined to scale the slippery pole to the heights of professional success. Although not directly relevant to the subject of why organizations that acquire knowledge-based services often seem unable to break free from the influence of external service providers, it seems anecdotally true that when in senior positions, where big decisions are made and large sums of money spent, managers begin to feel that they are invulnerable, or that they are just too astute to make bad decisions. Sadly they are wrong. It is a useful life skill, as well as a business or professional skill, to be able to learn how to buy knowledge effectively, and to ensure that knowledge, once bought, is transferred and utilized to the benefit of the organization. In short, to ensure that value for money is delivered.

It is hoped that this book will go some way to assist organizations to successfully 'buy knowledge'. The book is divided into a number of related subjects, all with the

1 Roderick M. Kramer, 'The harder they fall', *Harvard Business Review*, October 2003.

unifying theme of buying knowledge. We look at the knowledge economy, to set the scene on the manager's growing responsibility to buy in knowledge for their organization. We look at intellectual property rights and how they are created, transferred and protected. We set out some alternative strategies to *buying* knowledge. Working with universities, contract research organizations and consultancy firms is also explored. The most neglected area of all – knowledge transfer from 'supplier' to 'buyer', is given an exhaustive treatment.

This book is meant to encourage a review of your present methods, so you can benefit quickly by working through the text and apply learning points in your own context.

Wherever you see this symbol, it flags a useful learning point for client organizations. We suggest you review these in the context of your organization and determine any action necessary.

Throughout this book we refer to the 'buyer' of knowledge as the client, or client organization. This is to draw out a distinction between the buying function, which often places an order or a contract, and the internal customer who needs – and probably pays for – the service to be delivered by the knowledge 'seller'. We further refer to the client's 'organization' rather than 'company', in recognition that many buyers of knowledge do not work in industry or commerce, or even in the private sector. Client organizations that are involved in buying knowledge cover every aspect of business life. The lessons in this book, therefore, are universally applicable.

CHAPTER 1

In the Know – The Knowledge Economy in the Twenty-first Century

A BRIEF HISTORY OF KNOWLEDGE

First of all, what is knowledge? This might seem an extraordinary question, but it is vital to be certain what we are discussing. When managers invest their organization's funds to buy-in new knowledge, skills or technology, do they always fully recognize the element of *knowledge creation* – and the need to *acquire* that knowledge for their company? Collectively, do we recognize within organizations the value of knowledge?

The UK's *Concise Oxford Dictionary* is disappointing in its definition of knowledge: 'a theoretical or practical understanding of a subject, language etc., the sum of what is known ... certain understanding, as opposed to opinion'. For the purposes of this book, knowledge means accumulated experience, data and understanding gained by practical enquiry, reduced to useful applications. Furthermore, in the context of this book, with its emphasis on buying knowledge, we might add that knowledge contributes to the measurable 'intellectual capital' of organizations.

The heading above, *a brief history of knowledge*, is obviously tongue in cheek. What follows here is an admittedly selective canter through the major areas of human knowledge and recent trends in organizational development, as they relate to knowledge. This is simply to alert us to the breadth of knowledge now at the disposal of mankind.

The *Encyclopaedia Britannica*, 11th edition, categorized knowledge into 24 areas, alphabetically arranged. Although now considered rather passé by *Encyclopaedia Britannica*, which in later editions reassembled knowledge into ten basic areas, the 11th edition categories serve our purpose well enough:

1 anthropology and ethnology

2 archaeology and antiquities

3 art

4 astronomy

5 biology

6 chemistry

7 economics and social science

8 education

9 engineering

10 geography

11 geology

12 history

13 industries, manufactures and occupations

14 language and writing

15 law and political science

16 literature

17 mathematics

18 medical science

19 military and naval

20 philosophy and psychology

21 physics

22 religion and theology

23 sports and pastimes

24 miscellaneous.

The 15th edition of the *Encyclopaedia* contains in its introduction a short article by Mortimer J. Adler about the philosophy of knowledge. In considering humankind's thirst for knowledge, Adler commented:

> Among the things that man seeks to know and understand is his own knowledge – his abilities, efforts and achievements in the sphere of knowing itself. Whether or not Aristotle was right in saying that the highest form of intellectual activity is thinking about thinking itself, it is certainly true that 'knowledge become self conscious' is a distinctive characteristic of the human enterprise of knowing. We not only seek to

> know whatever can be known, but we also, reflexively, turn our knowing back upon itself when we pay attention to how we know what we know, the various ways in which we know, and the divisions and branches of our knowledge.[1]

Senior and middle-ranking managers in industry, government, commerce, academe, the military and other enterprises interact with other organizations, colleagues, regulatory agencies and a range of 'stakeholders'. Sometimes the need for new information is competitively driven – the need to get ahead or stay ahead of the competition. Sometimes the need for new information is collective in its incentive, where information is freely shared in the expectation of collectively gaining more knowledge. Whatever the motivation, managers must recognize when a *requirement for new information* has emerged and not confuse it with other acquisition objectives, where the need for knowledge may only be a secondary consideration.

What are the pressures on organizations today to advance knowledge and to keep abreast of new knowledge in the knowledge economy? Some of these pressures are suggested in Figure 1. New technologies pull organizations to new levels of efficiency

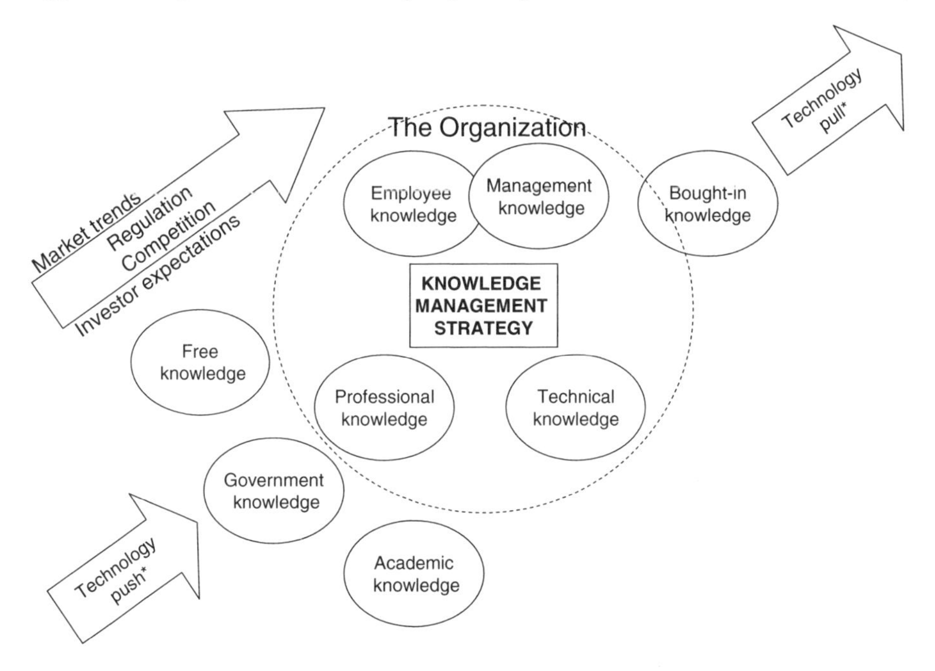

Figure 1 Organizations at work – pressures to advance knowledge

1 *Encyclopaedia Britannica*, 15th edition, p. 475, Propaedia. Knowledge become self conscious, Mortimer J. Adler.

and effectiveness; existing technologies push laggard organizations to catch-up. There are knowledge stakeholders both within and external to the organization. Markets and regulatory trends also push the organization to new areas of knowledge competency. Bought-in knowledge is another stimulus that can advance and quicken the transition to new knowledge competencies.

THE GROWTH OF KNOWLEDGE

We have already observed in the Introduction to this book that the twentieth century may, in retrospect, be viewed as *the century of exponential growth of knowledge.* It is worth adding that the rate of increase in knowledge accelerated dramatically in the last quarter of that century. The twenty-first century, no matter whatever else it may be marked by, will be characterized by accelerated growth in knowledge. The information age is here.

Can organizations really *buy knowledge*? Is not knowledge something that is acquired by diligent study and hard work? It is the author's view that not only can organizations buy and internalize knowledge, but that this will become an increasingly common element of organizational life in the near future, indeed a core competency for managers in the first decade of the twenty-first century. Knowledge-buying processes need to catch up with the reality of the information age. Already knowledge management (popularly known as KM) is a growing discipline of management. Knowledge buying will become a subset of KM. For the knowledge-based organization the core competency for survival is KM. In the knowledge-based economy at the beginning of the twenty-first century, KM is a critical element of strategy that allows organizations to accelerate the rate at which they handle new challenges and opportunities, by leveraging the key resources of collective know-how, talent and experience.

New knowledge abounds. For some organizations, new knowledge must be created specifically, but in a dynamic and constantly changing knowledge economy, the value of today's knowledge is of less and less duration. Knowledge has been likened to radioactive isotopes, in that its value decays at various rates based on its composition.

> Some knowledge has a half-life of days, while other forms of knowledge will endure for aeons. For example, the knowledge applied to popular fashion is clearly short lived. We all laugh at how silly we look in the clothes we thought so stylish in pictures taken just ten years ago. On the other hand, the knowledge gained from the great democratic experiments of ancient Athens still applies in our modern day forms of

government. Our notion of democracy has changed and will continue to change, but is still based upon and relevant to tenets that emanated centuries ago.[2]

If you are about to buy a knowledge-based service, how long will the knowledge be required to provide competitive advantage?

The difficulty in acquiring new knowledge can be a factor of how close the knowledge owner, broker or stakeholder is to your organization. The proximity of knowledge can be thought of as a series of concentric circles with the 'distance' of knowledge agents from your organization plotted as suggested in Figure 2. It may be beneficial to create a similar chart for your own organization, carefully thinking through how different knowledge agents interact and what sort of commercial relationship will best serve both parties' needs.

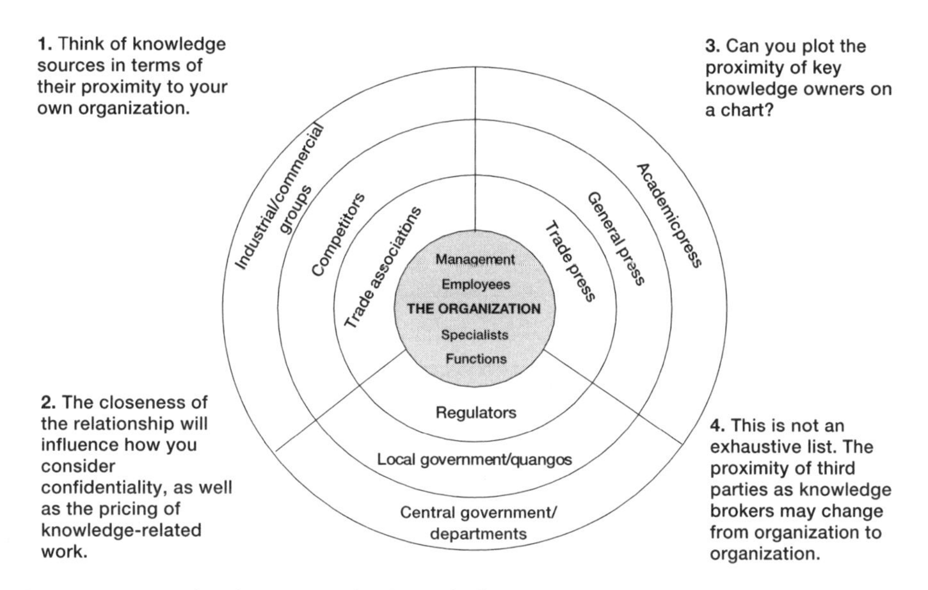

Figure 2 Organizations at work – knowledge proximity

TECHNOLOGICAL RESEARCH AND DEVELOPMENT

Plainly, not all new knowledge of significant value to organizations emerges from the field of science and technology. But it is probably true that significant *disruptive*

2 T. Koulopoulos and C. Frappaulo, *Smart Things to Know About Knowledge Management*, Capstone 1999, p. 15.

technologies[3] will emerge from the science and advanced technology sectors. So an understanding of science policy is useful background to the development of a KM strategy.

Accurate data on research and development (R&D) expenditures are notoriously difficult to obtain. The basis on which costs are allocated to research varies from company to company as well as from country to country. Even definitions of what is, and what is not research, do not have universal acceptance.

Many companies are reluctant to reveal the true extent of their R&D investment and can legally (within the accounting rules of most Organization for Economic Development [OECD] countries) apportion expenditures to disguise exactly where money is being spent. To overstate or accurately state R&D expenditure may invite unwelcome scrutiny by investors and competitors: to understate may reveal an unpalatable truth about the company's trading position. As with companies, so it is with countries that have differing accounting policies and, for political reasons, may also wish to reflect their technological prowess in a more favourable light.

It must be said, then, that the figures quoted even by such august bodies as the OECD may reveal only a distorted picture of R&D expenditures. The following table (Table 1) which shows the percentage increase in corporate R&D over the period 1992/96 and which is based on OECD figures, can however be taken as generally accurate.

Table 1 National corporate R&D spend: acceleration 1992–96

	1	*2*	*3*
Sweden	82	7.4 (4)	19.1
Canada	72	10.8 (2)	NA
Denmark	47	15.1 (1)	NA
Switzerland	40	6.2 (5)	15.4
US	32	4.3 (9)	20.9
UK	22	2.3 (12)	17.8
France	17	4.0 (10)	13.5
Netherlands	16	5.3 (6)	NA
Germany	13	4.7 (8)	17.2
Japan	3	4.9 (7)	56.5
Italy	–2	2.3 (12)	17.6
Belgium	–8	NA	NA

Column 1: Percentage increase in corporate R&D investment 1992–96 by the top 300 international companies.
Column 2: R&D/sales ratio 1996 (figures in parenthesis indicate national ranking).
Column 3: P/E ratio (1992–96 average).

3 New technologies that completely reshape an industry – a classic example in the UK was the invention of the bagless (Dyson) vacuum cleaner.

Table 1 indicates countries in which companies are accelerating R&D spend (column 1) together with their R&D/sales ratio (Column 2) which indicates the ratio of R&D expenditure to sales income. Finally the Price/Earnings (P/E) ratio in column 3 shows the average P/E ratio across all industry sectors. Whilst acknowledging the 'broad-brush' nature of these figures, they do reveal that in some countries commercial companies are investing heavily to strengthen their technology base. Traditionally strong countries (Germany and Japan) are temporarily slowed, but both have a massive advantage in R&D investment that is likely to be maintained. It may take until 2015 or longer for Germany to overcome the fiscal indigestion of unification. This may give her local competitors an opportunity to narrow the technology investment gap. A higher than average increase in investment by some countries may indicate a pre-existing investment gap that they are trying to close, or that they are targeting new technology opportunities.

For organizations considering sourcing knowledge-based services overseas, these figures may indicate real value-added opportunities in certain countries, especially if that country already has strengths in the particular field of interest. (In the author's direct experience, Eastern Europe offers value-added opportunities at the beginning of the twenty-first century.) Perversely, countries where investment has slowed or stagnated may provide opportunities for 'at cost' R&D as some of their R&D organizations look more actively for contract work to preserve their capabilities.

The most often quoted comparison of international expenditure on research and development is gross domestic expenditure on R&D (GERD)[4] as a percentage of GDP. GERD includes all expenditure on R&D, both public and private sector, civil and military. Recent figures are given in Table 2.

Useful publications for managers who need more data on international comparisons of R&D can be found in:

- The OECD publication: *Main Science and Technology Indicators*. This is published annually and is obtainable in the UK from the Stationery Office

4 Resources allocated to a country's R&D efforts are measured using two indicators, R&D expenditure and personnel. For R&D expenditure, the main aggregate used for international comparisons is gross domestic expernditure on R&D (GERD) which represents a country's domestic R&D-related expenditure for a given year. The R&D data are compiled on the basis of the OECD's *Frascati Manual* which defined research and development.

- The UK *R&D Scoreboard.* This is published in the UK annually and is obtainable free of charge from the UK Department of Trade and Industry. Also available on line at the DTI website: www/innovation.gov.uk

Table 2 Gross domestic expenditure on R&D as a percentage of GDP

Country	Percentage	OECD share
Canada	6.50	2.8
France	2.60	5.3
Germany	3.50	8.3
Italy	2.70	2.4
Japan	3.00	16.7
United Kingdom	2.50	4.2
United States	6.00	43.7
EU average	4.00	28.1
OECD average	4.50	100

Source: OECD (MSTI database) May 2003.

Global aggregate R&D expenditures are very large. The US alone spends approx $280 billion pa on civil and military R&D and the rest of the world is estimated to bring this figure in dollars equivalent to $645 billion. Estimating levels of externally sourced *contract research* is difficult, but assuming the figure is of the order of 10% for bought-out R&D services, in the sense of purchasing knowledge via specific, packaged work, the market for contract research may globally be worth $70 billion. Add to this the cost of bought-out goods and services and capital investment in facilities, then the annual total spent by researchers around the world is of the order of $100 billion to $120 billion.

The division of this expenditure between government and industry is indicated in Table 2, which shows an interesting level of parity between the US and the EU. Closer inspection reveals, throughout the Western bloc, that there has been a retreat by governments from R&D investment with the shortfall being taken up by industry. The difference between private and state funding for R&D is most pronounced in Japan, with industry funding the lion's share. This is all the more emphatic considering the comparative lack of military R&D by Japan and the fact that in the US military R&D accounts for 50% of all US R&D funding. This goes some way to explain the Japanese dominance in export revenue-earning advanced-technology consumer goods of all types. Reliable comparative data for the former Soviet bloc nations and China is not available.

Different industries invest in R&D at different rates: Table 3 compares the level of R&D as a percentage of sales for seven industry sectors. Aerospace, at nearly 20% of sales, is clearly the leader in this particular league. It does not, however, follow that contract research expenditure is similarly high. Aerospace must of necessity spend the

Table 3 R&D spend – government and industry share of total-trend 1981–93

Funding source	*Government*		*Industry*		*Other national sources*		*Abroad*	
	1981	*1993*	*1981*	*1993*	*1981*	*1993*	*1981*	*1993*
North America	49.3	39.6	48.4	57.6	2.0	2.3	–	–
Japan	24.9	19.6	67.7	73.4	7.3	7.0	0.1	0.1
European Union	46.7	39.7	48.7	53.2	1.1	1.4	3.5	3.7
OECD	45.0	36.2	51.2	58.8	2.4	2.9	–	–
Research undertaken by	*Government*		*Industry*		*Private non-profit*		*Higher education*	
	1981	*1993*	*1981*	*1993*	*1981*	*1993*	*1981*	*1993*
North America	12.6	10.8	69.3	70.3	3.0	3.2	15.1	15.7
Japan	12.0	10.0	66.0	71.1	4.5	4.9	17.6	14.0
European Union	18.9	16.5	62.4	62.6	1.4	1.4	17.4	19.5
OECD	15.0	12.7	65.8	67.4	2.6	2.9	16.6	17.1

bulk of its research and development funds in-house, although a high proportion of this will be in work force and other goods, materials, services and facilities types of expenditure. Similarly, pharmaceuticals also spend a high proportion in-house, but are increasing the extent of bought-in knowledge – otherwise known as contract research.

Table 4 Industry sector R&D investment as a percentage of sales

Aerospace	19.0
Pharmaceutical	14.0
Electronics	7.5
Chemical	4.0
Automotive	3.5
Mechanical engineering/optics	3.5
Energy	2.0

At the other end of the spectrum, the energy industries which invest a comparatively low level of R&D funding to secure sales – perhaps reflecting a lack of technological competition in these highly regulated industries – are big spenders on external research. This may also reflect, in the Western G7 countries in particular, the influence of the four 'Rs' – Restructuring, Right-sizing, Rationalization and Re-engineering, which has hit hard on R&D budgets throughout the energy sector and led to an increased emphasis on buying in research that even a few years ago would have been seen as a core capability and carried out in-house.

The place of contract research in the EU in terms of national shares of the overall market was estimated by Buy Research (Cambridge, England) in 2002 as being:

Germany	30%
United Kingdom	23%
Netherlands	16%
Italy	12%
France	13%
Others	6%

Considering the relative size of the R&D budgets for these nations, these comparative expenditures suggest that France and Italy under utilize outsourced R&D opportunities.

Table 5 Top 10 corporate R&D investors – national ranking 2001–03

2001/2		*2002/3*	
Ford Motor, US	4 552 1	Ford Motor, US	4 782 2
General Motors, US	4 418 2	Daimler Chrysler, Germany	3 957 1
Daimler Chrysler, Germany	3 982 7	Siemens, Germany	3 792 8
Siemens, Germany	3 515 1	General Motors, US	3 602 7
IBM, US	3 280 8	Pfizer, US	3 215 1
Matsushita Electric, Japan	3 080 7	Toyota Motor, Japan	3 084 6
Ericcson, Sweden	2 974 2	IBM, US	2 946 1
Motorola, US	2 970 2	GlaxoSmithKline, UK	2 936 0
Pfizer, US	2 968 9	Matsushita Electric, Japan	2 884 2
Cisco Systems, US	2 729 2	Volkswagen, Germany	2 849 0

Figures are in pounds sterling millions equivalent.

Table 6 Global top 700 corporate R&D investors: national ranking 2002–03

	2002/3	
	Number	*Total spend £(m)*
Australia	1	194
Belgium	3	478
Canada	7	1883
Denmark	7	709
France	33	12 942
Germany	46	23 983
Italy	5	2114
Japan	155	44 689
Norway	3	186
South Korea	6	2879
Spain	4	292
Sweden	13	3578
Switzerland	21	6375
Netherlands	8	4567
UK	37	10 483
US	340	86 283

Source: UK DTI. Note that a number of countries have been omitted. Figures above focus on the main R&D-spending nations.

The financial sums expended by the major corporate investors in R&D are impressive. The figures in Table 5 show the top ten R&D spenders in 2001 and 2002. It is notable that the expenditures for the group even at the bottom have increased dramatically.

Of the global top 700 corporate investors in R&D in 2002/3, the nation by nation ranking given in Table 6 shows in each year the number of representatives within those 700 (left column) and their aggregate spend (right column, in pounds sterling equivalent).

These corporate expenditure figures indicate that, by comparison with previous years' statistics, the US and Japan are increasing their lead. In the case of the US this is unsurprising, given the state of the burgeoning US economy. For Japan and South Korea, in severe recession over the period, the expenditures show considerable corporate commitment to technology. France and the UK have maintained their positions.

CHAPTER 2

Knowledge is Power

SCIENCE AND SCIENCE POLICY

A significant trend at the beginning of the twenty-first century is the increasing importance of multinational organizations in the formulation of science policy. Most research institutions continue to be national in character and the bulk of financial support for these institutions continues to be from government and other state agencies. Several trends however, including the transnational character of many scientific problems and the emergence of regional free-trade areas, were at the turn of the century eroding the purely national character of many such institutions and increasing interest in multinational policy approaches.

Governments and private companies in the US, the EU and Japan account for approximately 85% of the world's investment in and performance of research and development. Emerging, however, are several other economies – particularly in East Asia and the Pacific Rim – where R&D and trade in high technology is the most significant long term science policy trend at the beginning of the new century.

JAPAN

Japan's R&D investment, which had been decreasing in the early 1990s mainly due to a reduction in private sector investment, in turn caused by poor trading conditions and the initial effects of recession in the Japanese economy, showed an increase over the period 1997 to 2003. Japan's investment at 3.0% of GDP is still impressive with a roughly 80/20 split between the private and state sectors respectively. In spite of this lead, Japan remains concerned that its research system lacks breadth and flexibility. In July 1996, the Japanese Cabinet adopted the Science and Technology Basic Plan which was intended to 'promote science and technology policies comprehensively, systematically, and positively from a new standpoint ... and to provide concrete science and technology promotion for five years from 1996 to the 2000 fiscal year'.

The most important feature of this Basic Plan was the Government's commitment to double its 1992 R&D investment by the year 2000, making its share of such R&D comparable to that of the US and the EU. The Plan pledged to shift R&D expenditure toward the solution of global problems such as natural resource depletion, the environment generally, energy use and food shortages, but also emphasized the importance of basic research 'which produces the common intellectual property for human kind'. Latest OECD figures at the time of preparation of this book indicate that Japan, at 3% of GDP, still trails behind the EU average investment at 3.5%.

SOUTH KOREA

During the middle and late 1990s South Korean GDP grew by 8% compared to 3–4% average across the US, EU and Japan. This growth was fuelled by high-tech industrial success and in turn led to an increase in R&D. At the turn of the century R&D investment stands at 6.0% of GDP – ahead of France, UK and Germany. The private sector accounts for 70% of this investment – an industrial investment matched only by Japan. South Korea is moving aggressively to assert its aspiration to become a world leader in science – working hard with OECD and Asia-Pacific Economic Cooperation (APEC) multinational organizations and hosting the second annual meeting of the APEC ministers of science. Like Japan, South Korea intends to increase basic science and to create an institute for advanced sciences modelled on the US Institute for Advanced Studies in Princeton, New Jersey.

EUROPEAN UNION

Germany, France, UK and Sweden all invest more than 2.5% of GDP in R&D whilst Denmark, Finland, the Netherlands and Italy invest rather more than 1.5%. Western Europe continues to boast world-class R&D facilities across all areas of science and engineering and basic research continues to flourish in spite of pressure from most EU governments to deliver more short-term gains from investment. Germany slipped behind France in 1996 in its state-sector R&D investment, largely due to continuing high costs associated with unification. At the time of writing this book, however, there is consternation amongst French state-sector research institutions at the recent cutback in expenditure on basic science. Whether this is a short-term phenomenon remains to be seen. Unusually for the UK the position in science investment appears to be improving. The Chancellor of the Exchequer in March 2004 unveiled plans for a ten-year investment strategy for science, linking state expenditure with incentives for corporate investment.

Personnel issues continue to be an important aspect of EU science policy. European universities are struggling with sharply increased enrolments encouraged by negligible tuition fees and continuing high unemployment (a process known as 'massification') and the implications of this in terms of quality of graduates and effect on research are as yet unclear.

EU institutional research and development budgets continue to hold up in the early 2000s, whilst those of research facilities not under direct EU control such as the European Laboratory for Particle Physics (CERN) and the European Synchroton Radiation Facility (ESRF) face some uncertainty. Unlike the US, where most basic research is supported by federal funding and performed in university facilities, most EU governments continue to support a parallel system of basic research institutions in addition to universities. Examples include the Max Planck institutes in Germany and the National Centre for Scientific Research laboratories in France.

The EU's Framework Programme for R&D which provides partial, usually matching, support for pre-competitive applied research projects undertaken in two or more EU countries, is being used as an explicit instrument of European integration. Although its budget is less than 5% of the combined R&D budgets of the EU member states, the Framework Programme exerts a significant influence on national priorities. For example the fact that it restricts support to direct research costs as opposed to salaries, permits it to play an important catalytic role. Finally, the weakness of the euro has adversely impacted the spending power of EU research institutions. It remains to be seen how this particular problem will be managed.

THE US

In spite of budget difficulties in the US, federal R&D budgets have been upheld to a large extent with most major recipients of federal investment seeing small increases. Viewed in aggregate, US expenditure in R&D is healthy and competitive, totalling $282 billion, matching the combined totals for Japan and the principal EU players (although in non-defence R&D the US's combined direct competitors have a significant lead). The share of GDP, at 6% in 2003, increased from 1993 when it was approximately 3%. Recent figures on corporate R&D show a surge in investment, certainly among the bigger US corporations. The war on terror – a significant political factor in the early 2000s – has seen a significant shift towards military R&D. The long-term effect of this shift on non-military industrial research has yet to be assessed.

In the early 1990s industry and government accounted for roughly equal shares of R&D investment in the US. By 1997, however, the industrial share had reached 59% and the federal share had slipped to 36%. This disparity will probably increase in the early years of the twenty-first century as government retreats from traditional areas of support. To what extent this may impact US competitiveness in global markets remains to be seen.

An important federal-funded report into how well the US capitalizes on its R&D investments (*Capitalizing on Investments in Science and Technology*, published jointly by the National Academy of Sciences, National Academy of Engineering and Institute of Medicine in 1999), noted that large companies which had traditionally supported important long-term R&D were forced to reduce expenditure in this area. The report commented on the growing importance of external sources of R&D across most industrial sectors.

Many have reduced their research organizations, have learned to acquire ideas and technology from outside the firm, and have adjusted their sights towards nearer-term goals. Large manufacturers are giving suppliers far greater responsibility for engineering and design work, and some medium-size firms that specialize in particular technologies are emerging as the key sources of innovation.

In the aircraft industry ... long-term research has traditionally been funded by government, and with tight funding, companies are focusing R&D spending on short-term research and product development. The leading US manufacturer, Boeing, has drawn on its component suppliers for R&D. Suppliers, in turn, are outsourcing more R&D to their subcontractors.

In the auto industry ... the Big Three automakers rely more on suppliers, co-operating with each other through the United States Council for Automotive Research consortium, and increasing interactions with the government, especially DoE, through the Partnership for a New Generation of Vehicles.

Across the spectrum of industries, major corporations have reduced, sold or closed their research facilities. During the early and mid 1990s, IBM cut and refocused research spending. RCA's Sarnoff Research Centre, the source of pioneering research in video, liquid crystals, lasers and other fields, was spun-off to SRI International following GE's acquisition of RCA, and converted to contract research.[1]

Organizations investing seriously in technological or scientific research should monitor science policy in the various trading blocs.

FUTURE PROOF – THE NEED TO MONITOR KNOWLEDGE

New technologies, the advance of science, increasing regulation and changes in society at large make it essential that organizations monitor developments generally, in order to be able to develop and reorient themselves, and respond to new and impending developments. It is unlikely that any major organization today will contain or be able realistically to develop all the information and knowledge that they need. A simple illustration: in the early days of aviation the technologies, skills and knowledge required to develop, build and operate aircraft were capable of containment within a

1 *Capitalizing on Investments in Science and Technology*, NAS, NAE, IOM, 1999, pp. 35–6.

single organization. So, motor engineers were able to branch into aviation engineering with little difficuty – the engineering skills and knowledge were similar. As technology developed, however, aircraft firms were required to absorb a greater range of skills – for example stress analysis, metallurgy, fluid dynamics, ergonomics and psychology. Some firms were able to bring major elements of these new disciplines in-house. But as technology advanced further into advanced electronics, new sources of power and fuel and computer-controlled flight dynamics, aero industry management were required to make key strategic decisions. Rather than bring all skills in-house, the industry developed a pool of specialist suppliers with expertise in areas that were vitally important technologies, but of secondary importance to the core competencies of design, development, integration and manufacturing.

The aviation experience can be seen in other industries. As customer demands grow, products become more complex, ever greater reliability and effectiveness are required, and self- or government-regulation increases. Organizations have to make strategic choices about what knowledge and skills are required in-house, and what can safely be outsourced. It is becoming increasingly difficult, in any case, for organizations to remain abreast of all developments.

Globalization has forced organizations to become more effective in developing new products and processes as increasingly fierce competition threatens virtually all markets. Technology is becoming ever more complex, product life cycles are shortening and substitute technologies follow in rapid succession. This is true of every industry and every activity that has a technology/knowledge basis. Organizations are finding it increasingly difficult to carry the internal resources to sustain knowledge capabilities in all areas of interest and therefore concentrate capabilities into core competencies. Knowledge management in other non-core capabilities is increasingly managed through forming strategic partnerships or contracting out non-core activity.

Buying knowledge services is a form of commercial and organizational integration of the buyer and the seller, albeit on a temporary basis. The decision to purchase knowledge services should be based on the overall strategy of the organization, but evidence shows that this may not always be taken in a systematic fashion. Readers who need to explore the problems, pitfalls and rationale of knowledge procurement in advanced technology/science industries can pick up useful information in *The Outsourcing R&D Toolkit* – full details of which are contained in Appendix 1.

KNOWLEDGE IS POWER: CREATING AND DIFFUSING NEW KNOWLEDGE

It is no surprise that the world's wealthy nations – and for those read member countries of the Organization for Economic Cooperation and Development (OECD) – anxiously monitor their own and their direct competitors' ability to create and benefit from new knowledge. Statistics can be dry, but sometimes reveal very important trends – trends that if followed through could lead to a wholesale change in a particular situation. Just such a change may be on the way for the rich nations which currently control the bulk emerging new knowledge. According to the OECD's R&D productivity data for 2003 the US, Canada, the Netherlands and Australia, amongst the OECD nations, received the largest boost from investment in information and communications technology (ICT). Much of OECD labour productivity growth in 2003 was concentrated in knowledge intensive activities, notably ICT services, together with high-technology and medium-high technology manufacturing.

A significant new development in the first few years of the twenty-first century was that the ability to create and use new knowledge through investment in technological and scientific R&D, use of ICT, development of scientists/engineers and the filing of patents, was extending to a wider range of countries – many of them outside the OECD's membership. This suggests increasing competition for the *factors of knowledge creation* – skilled people, innovative businesses and capital – with a likely reduction in some of the broad advantages that OECD countries enjoyed in the 1990s.

OECD-wide investment in R&D rose in 2001 and into 2002, while patenting nearly doubled over the decade. This was stimulated by developments in the ICT and biotechnology sectors. This activity however is no longer the sole province of the OECD countries. Major non-OECD economies currently account for 17% of global R&D expenditure, with Chinese R&D expenditure of some US $60 billion, putting China third in the world behind the US and Japan. India spent about US $19 billion on R&D in 2000–01, putting it among the top 10 countries worldwide. Chinese Taipei was the fourth largest recipient of US patents, ahead of France, the UK, Korea and Canada.

Human capital is an essential factor of economic growth based around science and technology. In the early 2000s universities in the EU awarded 36% of science and engineering degrees in the OECD area while the US universities awarded 24%. To compensate, the US continues to draw on the skills of foreign-born scientists and engineers – a continuation of Europe's perennial complaint of a 'brain drain' in favour of the US – first recognized in the late 1950s. While some OECD countries such as the UK and Canada are important sources for scientific personnel in the US, today three times as many foreign-born scientists are from China and twice as many from India as from the UK. In many cases, these foreign-born workers come from the national university system. Foreign students represent more than a third of Ph.D. enrolments in Switzerland, Belgium and the UK, 27% in the US, 21% in Australia, 18% in Denmark and 17% in Canada. In absolute numbers, the US has far more foreign Ph.D. students than other OECD countries, with around 79 000 in 2003. The UK in the same year followed with some 25 000.

ICT continued to spread, despite the slowdown in parts of the sector. In Germany, Sweden, Denmark, and Switzerland some two-thirds of households had access to a home computer in 2002. In many OECD countries 80% or more of commercial enterprises with ten or more employees use the Internet – and this includes countries like the Czech Republic and Spain. Broadband access is more varied and widely diffused in the US, Korea, Canada, Sweden, Denmark and Belgium. In Sweden and Denmark, 20% of commercial companies have access to the Internet through a connection faster than 2Mbps. The integration of the Internet into everyday life continues at a rapid pace. In the US, almost 40% of Internet users buy online. The share of electronic sales in total US sales grew by 70% between 2000 and 2002, reaching 1.5% of retail sales. In Sweden and Portugal about half of all Internet users play games online and/or download games and music. In Sweden and Denmark, more than 50% of Internet users undertake e-banking.

KNOWLEDGE IS POWER: PRODUCTIVITY AND ECONOMIC STRUCTURE

The growing *knowledge intensity* of the OECD economies has been accompanied by rapid economic globalization. The trade to GDP ratio increased by about two percentage points over the 1990s in the US and EU, although it remained stable in Japan. Trade in high technology goods, such as aircraft, computers, pharmaceuticals and scientific instruments, accounted in the early 2000s for over 25% of total trade, up from less than 20% in the early 1990s. A significant fraction of this trade was between different affiliates of multinational enterprises. The share of intra-firm exports in total exports of manufacturing affiliates under foreign control ranged between 35% and 60% in OECD countries at the beginning of the twenty-first century.

The amount of manufacturing R&D expenditure under foreign control grew by nearly 90% between 1993 and 1999 (at current 2003 prices) with the US being the destination for nearly half of this investment, accounting for about 18% of all US manufacturing R&D in 1999. For many countries, including the UK, the Netherlands, Spain, Sweden, Canada, Ireland and Hungary, foreign affiliates account for 30%, or more, of manufacturing R&D. In Ireland the figure is 70%.

Some OECD countries, thanks to a combination of factors, increased growth during the 1990s. These factors included higher labour utilization, capital deepening – notably in ICT, and more rapid multifactor productivity (MFP) growth. Over the second half of the 1990s, MFP growth accounted for a considerable part of overall growth of GDP, particularly in Finland, Greece, Ireland and Portugal. By 2000, services accounted for 70% of OECD GDP; manufactures accounted for about 18%. In many OECD countries, business services currently account for the bulk of labour productivity growth. Part of the increase in the service sector's contribution to value added reflects the manufacturing sector's greater demand for services, some of which is due to the outsourcing of activities previously undertaken in-house. Unsurprisingly, estimates of the amount of services embodied in one unit of final demand for manufactured goods show that it was significantly higher in the mid-1990s than in the early 1970s.

Investment in knowledge in the OECD is defined as the sum of R&D expenditure, expenditure for higher education (public and private) and investment in software. In 2000 investment in knowledge amounted to 4.8% of GDP in the OECD area, and would be around 10% if expenditure for all levels of education were included in the definition. The ratio of investment in knowledge to GDP is 2.8 percentage points higher in the US than in the EU. In Sweden (7.2%), the US (6.8%) and Finland (6.2%) investment in knowledge exceeds 6% of GDP. In contrast, it is less than 2.5% of GDP in southern and central European countries and in Mexico.

In the early 2000s most OECD countries were increasing investment in their knowledge base. During the 1990s, investment increased by more than 7.5% annually in Ireland, Sweden, Finland and Denmark – far above the increase in gross fixed capital formation. The amount of investment in knowledge was still low in Greece, Iceland and Portugal, although growth of GDP was similar to that of most of the knowledge-based economies (such as Sweden and Finland). In the US, Australia and Canada, gross fixed capital formation grew more rapidly than investment in knowledge.

For most countries, increases in software expenditure proved to be the major source of increased investment in knowledge. Notable exceptions were Finland (where R&D was the main source of increase – taking Finland to the second position in the R&D-intensity table published by the OECD, behind Sweden) and Sweden (where all three components grew). Gross fixed capital formation also covers investment in structures and machinery and equipment, which is a channel for diffusing new technology, especially to manufacturing industries. Gross fixed capital formation accounts for around 21.3% of OECD-wide GDP, of which machinery and equipment accounts for around 8.4%. The ratio of investment in machinery and equipment to GDP varies from 6% (Finland) to 14.6% (Czech Republic).

OECD-area R&D expenditure continued to increase steadily in recent years, rising by 4.7% annually between 1995 and 2001. Since 1995, growth in the US (5.4% a year) outpaced growth in the EU (3.7%) and Japan (2.8%). In 2001, R&D expenditure in the US accounted for approximately 44% of the OECD total, close to the combined total of the EU (28%) and Japan (17%). Below-average growth in R&D expenditure in the EU was mainly due to slow and declining growth of the major European economies. Compared to average growth in the OECD area over 1995–2001 (4.7%), R&D expenditure increased by only 3.2% a year in Germany and by less than 3% in France, Italy and the UK.

In the three main OECD regions – the US, Japan and the EU, R&D expenditure relative to GDP (R&D intensity) continued to increase steadily over the first three years of the twenty-first century. In Japan, this was due more to the stagnation in GDP since 1997 than to significant increase in R&D expenditure. In the US, however, the rise was mainly due to significant increases in R&D expenditure, as GDP also grew rapidly. In 2001, R&D intensity in the EU exceeded 1.9% for the first time in a decade. In 2001, Sweden, Finland, Japan and Iceland were the only four OECD countries in which the R&D-to-GDP ratio exceeded 3%, well above the OECD average of 2.3%.

KNOWLEDGE IS POWER: NEW ENTRANTS IN THE GLOBAL KNOWLEDGE STAKES

The somewhat dry statistics in the preceding section show the concern and the determination of the wealthy nations to compete in the global knowledge marketplace. But powerful new competitors are emerging to challenge the dominance of the OECD nations – a state of affairs that allows intriguing new possibilities for buyers of knowledge and knowledge-based services. Mighty US combine General Electric (GE) is today leveraging the global skills base. Until the year 2000 almost all of GE's spending on research, as opposed to product and process development, took place in the US. But following a corporate decision to become less focused on the US, GE has opened important new research facilities in Bangalore, Shanghai and Munich. In taking this decision, GE followed a trend set by Siemens of Germany, Philips of the Netherlands and IBM – the latter widely considered to be the leader in global research, with some eight laboratories, of which just three are in the US.

By spreading their research base these corporate giants hope to access a wider range of technological competencies than may be available (at an affordable price) in their home countries. GE's vice president in charge of R&D, Scott Donnelly, was tasked in 2000 with overseeing the investment of $280 million to build new research facilities globally. Donnelly's philosophy is that, thanks to GE's remarkable breadth of activities (power stations to X-ray scanners, consumer products to financial services) a number of critical research technologies can be 'leveraged' across GE's eleven main business divisions. His four main research business units were organized into globally transparent cross-functional teams. Different people work on the same technologies in different parts of the world, connected by e-mail and in frequent contact. The research teams focus on a dozen 'core' or 'enabling' technologies, covering areas such as solid state physics, organic chemistry, electronics and imaging, metallurgy and bio sciences. GE's advantage, according to Donnelly, is that it can often use the same idea in a core/enabling technology in many ways. A further challenge is to keep the research teams connected closely to the development teams within the business units. New knowledge can then be delivered rapidly in products to the market.

Whilst many companies are now establishing research and other knowledge-focused operations around the globe, even more are beginning to take advantage of the growing pools of talent that can be found and nurtured in far-flung places in the world. Research collaborations have for decades been a valuable mechanism for stimulating the development of new knowledge. But the emerging markets are also becoming emerging knowledge powerhouses. India is perhaps the prime example. The idea that all that India has to offer is cheap labour and a telecommunications link – the two factors that have made it a favourite destination for outsourcing of operations – is now out of date. Today India's knowledge workers are involved at the high end of technological and scientific research and development.

India has not yet sought to draw attention to its growing technological competencies. Conscious of Western fears of the migration of ever more technically demanding jobs, some have concluded it is best not to draw attention to themselves for fear of the growing resentment and determination of Western labour, unions and businesses to preserve capabilities at home. A veil of discretion currently masks some of India's R&D achievements. Some multinationals such as IBM are loath to discuss the subject at all. Others, such as Texas Instruments, boast that they have 'increased design resources around the world' but add that this has been achieved without transferring significant numbers of jobs abroad. Some Indians are themselves sceptical about their ability to compete in the truly big league, noting not only a lack of an in-depth scientific and technological infrastructure, but also weaknesses in the ability to create and protect intellectual property. Yet by crude measure of patents filed, India does indeed have strengths that should be of interest to knowledge buyers. In recent years major Western names such as Intel, Oracle, Texas Instruments, Cisco, GE, ICI, Whirlpool and SAP have together filed something in excess of 800 patents in India.

Monitor the capabilities of emerging markets. The range of tasks that can be outsourced is growing steadily.

MADE IN JAPAN: A KNOWLEDGE ECONOMY ADJUSTS TO LOW-COST RIVALS

In the late 1990s and early 2000s Japan's manufacturing moved steadily overseas. The principle reasons were to exploit lower wage economies – especially China – where wages can be 20 or 30 times less than in Japan. Concerned about the long-term implications of this shift, however, many manufacturing firms are devising strategies to fend off low-cost competition from overseas. Household names such as Canon and Toyota have developed integrated manufacturing systems that are more complex and sophisticated than low-cost, low-tech rivals can achieve. In other sectors, companies are learning how to protect skills and trade secrets that have for many years been key to their success. Others are even bringing back to Japan (so-called 'insourcing') activities that might otherwise be lost through knowledge leakage to rivals. These moves have received some encouragement from the Japanese Ministry of Economy, Trade and Industry (METI) which actively negotiates with Japanese manufacturers in an attempt to stem the leakage of jobs and technologies overseas.

Cheap labour may not be the sole reason for moving to other counties. Carmakers and machinery constructors have set up overseas simply to follow customer demand. Japan's second car maker, Honda, has commented that by building factories close to their markets, it can shorten production times, reduce exposure to currency fluctuations and lower distribution costs. Overseas R&D facilities can help to tailor vehicles to local market needs, such as sports-utility vehicles in America with extra towing capacity, a feature rarely needed in Japan.

Some manufacturers respond to competition by keeping secret their core technologies and core skills, whilst still shifting low value-added assembly and production activities to lower-cost destinations. Toshiba, Japan's biggest chip maker, is today moving ahead fast on developments of very advanced microchips and is avoiding past mistakes, where in cooperating with South Korean chip makers on basic chip technology, it effectively 'sold the family silver' only to find that former partners became tomorrow's competitors. Other manufacturers are finding that insourcing (bringing tasks that have been outsourced back in-house) can raise revenues and profits. This has highlighted three key advantages to the Japanese approach to manufacturing:

- well-trained, multiskilled employees
- low defect rates
- lean manufacturing processes that boost production flexibility and reduce inventory costs.

A 'white goods' manufacturer, Kenwood, has noted that its Malaysian workers, who have a high turnover, do not match up to its domestic Japanese workers, who remain employees for longer and so master a number of tasks – four or five per employee as opposed to one per employee on the Malaysian production lines. Consequently defect rates are anything up to 80% lower in Japan, and production times for equivalent products are markedly shorter. In the first few years of the twenty-first century the survival strategy of Japanese manufacturers revolves around industry's ability to identify, maintain and improve their inherent strengths. Professor Takahiro Fujimato of Tokyo University, in an influential article 'A twenty-first century strategy for Japanese manufacturing' (*Bungei Shunju*, November 2003), commented that it is overly simplistic to suggest that Japan should concentrate on high value-added production. He argued that what Japanese manufacturers really excel at is 'products whose functions require many components to be designed in careful detail and mutually adjusted for optimal performance'. Building such products required close teamwork, as well as cooperation with suppliers. Both elements reflect the need to manage knowledge components – the knowledge of workers as well as the knowledge and skills of suppliers that are bought in to client organizations.

The ability to combine different technologies and skills remains a key feature of Japanese manufacturing. As will be discussed in Chapter 5 (in the section *Is external technology acquisition increasing?*) companies and organizations generally are progressively becoming 'multitechnology' and their products are mul-tech (for multitechnology) as much as high-tech. For example, successful photocopier companies, such as Ricoh and Canon, combine advanced chemical processing for toner inks, precision mechanics and excellent servicing skills. Automobiles also require the synthesis of many technologies and knowledge-based skills. Nowadays it is difficult to think in terms of core technologies in autos, where there are a range of different technologies working together – ceramics, hydraulic, mechanical engineering, combustion engineering, metallurgy, electrical, computing and environmental, to name a few.

The rapid transfer of manufacturing by Japanese firms to South Korea, Taiwan and latterly mainland China, at the turn of the twenty-first century, forced many Japanese firms, as well as their government advisors, to consider at a strategic level which knowledge-based skills and technologies to protect, and which should remain based in Japan itself. As much of Japan's technological 'edge' depends on closely guarded trade secrets, this is becoming ever more complex in the early years of the new century. Often specialist suppliers will work with rival firms, and in south east Asia there is some clear evidence that low-cost rivals frequently 'reverse engineer' (the posh word for copy) their rivals' technologies and skills. Television manufacturer Sharp responded to this danger by trying to maintain specialist manufacturing equipment using in-house teams rather than utilizing the machinery's own manufacturer, simply to keep such manufacturers ignorant of potential defects in machines they may also have sold to direct competitors.

Where is your organization vulnerable to loss of information that could be of value to competitors, rivals or downright enemies?

Monitor information to be bought in to your organization for its potential for loss or leakage to others. Would it be better to develop new knowledge in-house, rather than share it with other organizations?

KNOWLEDGE IS POWER: TOWARDS A STRATEGY FOR INCREASING INNOVATION

Organizations base much of their strategy planning on the not unreasonable premise that innovation is an absolute basic requirement in the modern economy, and in modern society: innovation is an essential element in the constant struggle for competitive advantage for companies. Innovation enables society at large to enjoy greater benefits in public service delivery. But what is innovation? The *Oxford English Dictionary* describes innovation as 'making changes to something established'. Invention, by contrast, is defined as 'coming upon or finding: discovery'. Innovators work to change the status quo whereas inventors create (or stumble across) new things. In terms of buying new knowledge, both activities are potentially buyable, but it is naturally in the area of innovation – of making existing assets work more effectively – that most knowledge buying will be concentrated.

Whereas the benefits of invention may be immediately obvious, the advantages brought by innovation can often require greater management support. Changing the status quo is always uncomfortable. A certain household name PC manufacturer set out to imitate Dell Computer's world-famous 'build-to-order' system of computer assembly. The company found that its attempts were frustrated not just by its head of marketing, who did not want to disturb existing retail arrangements, but also by the head of manufacturing, who feared the innovation would lead to the outsourcing of most of his department's activity – and probably his own job. Both managers fought a powerful rearguard: at the time of writing this book Dell's dominance of this type of manufacturing continues without serious challenge.

The Gillette company utilized the advertising strap-line 'Innovation is Gillette' in the early 2000s, and most large companies would like to think of themselves as natural innovators. To such organizations innovation can seem like a silver bullet. Inject innovation into the right corporate/organizational areas and success will surely follow. It is rarely so simple, however. Blockbuster inventions are more and more difficult to create. Big companies can derive greater benefit from making lots of small improvements in existing products and processes – the essence of innovation. Even in the highly inventive business of pharmaceuticals, genuine new products are harder to find. Global spending on pharmaceutical research doubled in the decade to 2004, but the number of new drugs approved by the world's regulatory health agencies halved in the same period. 'Big Pharma' live in the hope – and often direct their efforts towards – finding a dramatic new drug, but many new and promising developments are bottled up and stored away in the laboratories because they cannot be profitably produced (and that means, in essence, passing all the essential regulatory hurdles).

'Disruptive innovation' – simpler, cheaper, better or more convenient products that seriously upset the status quo, can lead to the ending of the dominance of market-dominating players. The Dyson vacuum cleaner is a well-known example. Others are IBM's exposure once the PC had been invented, and the emergence of low-cost carriers in the airline business. Most organizations today grow through sustaining and directing innovation. So how do organizations stimulate innovation? And should this emerge from within or be purchased from externals?

Big companies have learned hard lessons from the history of innovation. Many have cut back and redirected investment in research. The old idea of ranks of white-coated boffins dreaming the future of the organization is now largely obsolete, although 'Big Pharma' and Microsoft are exceptions to the rule. In some industries reductions in R&D spending reflect changes in the way products travel down the invention pipeline. During the late 1990s, for example, Cisco systems of the US maintained its technological dominance of the high-tech Internet/server business by buying a succession creative start-up companies that had originally been financed by venture capital. The company's R&D was, it might be argued, outsourced to California's venture capital market, who combined the innovative flair of small corporations with the marketing know-how of large ones. We have already noted there are serious moves to 'outsource' research to emerging economies. Acquiring controlling stakes, or simply buying-up innovative firms, is another aspect of the same general move. This might be characterized as a fundamental change in the supply chain of invention. For example, in the biotech industry many firms act as intermediaries between universities and 'Big Pharma', sometimes with personnel moving between these firms as innovations come to maturity. Universities once licensed inventions direct to the big manufacturers, but small intermediaries can make the whole process more efficient.

Market fragmentation leads towards greater development of niches with the overriding requirement to customize products for smaller groups of consumers. This is true not only in commerce but in the delivery of public services, where the idea of 'choice' has become a (Western) politicians' mantra. So how do organizations gear up to improve and increase innovation? The following are key lessons:

- Reward and thereby encourage innovation among employees.
- Monitor market and other developments, whether technological or social.
- Monitor competitors or peer-group organizations.
- When buying new knowledge assess whether such knowledge will support innovation along with other organizational objectives.
- Companies should avoid the idea it is not worth investing unless the rewards are enormous. A series of smaller innovations might collectively lead to great rewards.
- Companies need to innovate constantly to defeat the 'commoditization' of products and services. They still need to find a unique selling point that encourages consumers/customers to recognize their product over their rivals'.
- Big manufacturers should consider hiving-off promising new products into independent business units, away from the smothering influence of the status quo.
- Big organizations should themselves attempt to become disruptive innovators.

CHAPTER 3

Head Knowledge – Modern Intellectual Property Rights

WHAT EXACTLY ARE IPRs?

Intellectual property (IP) law is a major discipline within the law. The treatment in this chapter is only intended as an overview and an opportunity to explore some practical issues that will emerge for those involved in buying knowledge. Accordingly the treatment is necessarily brief and only intended to alert readers to potential pitfalls and opportunities. Specific legal advice may be necessary to determine the optimum course to adopt in particular circumstances. Intellectual property has been described as:

> All those things that emanate from the exercise of the human brain, such as ideas, inventions, poems, designs, microcomputers and Mickey Mouse. ... The legal description of intellectual property differs from the colloquial in that it focusses on the rights which are enjoyed in the produce of the mind, rather than on the produce itself. In legal terms we call a piece of land, or a painting, or a motor car 'property' ... because individuals or legal entities such as companies can assert a right in it against some or all other persons. The word 'property' comes from the latin *propirus*, which means 'one's own'. If we bear this in mind, we can take the expression 'intellectual property' to mean the legal rights which may be asserted in respect of the product of the human intellect.[1]

In today's global economic context, where knowledge creation may take place under differing legal jurisdictions, it may well be necessary to take legal advice about specifics. From the client's point of view, with regard to bought-in knowledge there are two conditions to be achieved:

- free and unfettered use of the knowledge is obtained
- knowledge does not find its way into the wrong hands – for example, a competitor's!

---

1 J. Phillips and A. Firth, *Introduction to Intellectual Property Law*, 2nd edn, pp. 3–4, Butterworths, 1990.

The traditional method of protecting IP is to grant its owner protection against infringement by others of an exclusionary (often exclusive) right of exploitation for a limited duration. The owner knows once he has been granted an exclusionary right over his IP that his exploitation of it will be free from unwanted competition (which would be theft of his intellectual property right) or he will have legal redress against that competition.

The acquirer of an intellectual property right (IPR), either by developing the IPR himself or by buying it from a third party, can derive no benefit from his ownership except by exploiting it commercially. Patent or other IP ownership allows its owner to derive profit from ownership. Consumers benefit from being able to buy and use the IP product, the nation's work force gain useful employment from making or distributing the product and the government derives tax revenue from every stage in the process! All of which represents a very virtuous circle and one that helps to explain why the State is willing to grant temporary monopoly rights in a society that is otherwise not well disposed towards monopolies.

Considering intellectual property from the perspective of the organization and its need to acquire and manage knowledge, it will be seen that there are a range of knowledge sources, as suggested in Figure 3. Each of these sources may also be owners or brokers of intellectual property, so strategies for the acquisition and management of the attendant legal 'rights' will have to be devised by knowledge client organizations.

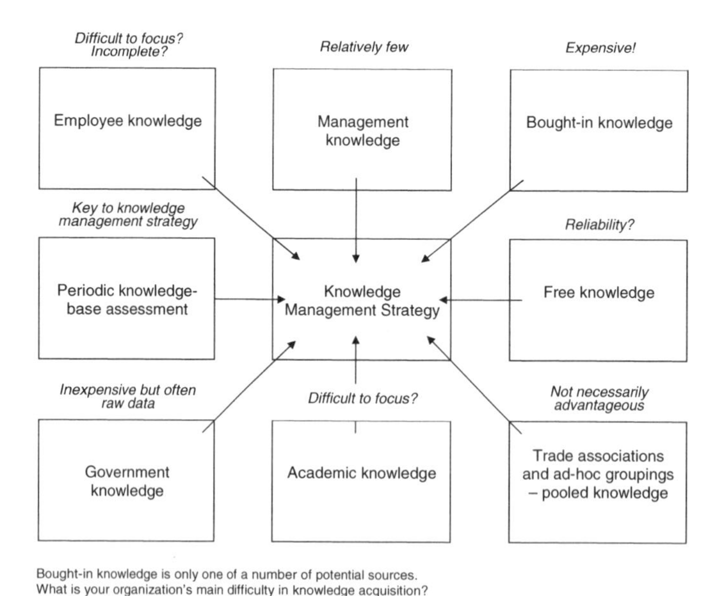

Figure 3 Organizations at work – the knowledge environment

PRACTICAL PROTECTION FOR IPRs

The brief descriptions that follow represent the situation in the UK (except in the case of utility models). Equivalent protection is given in most developed countries, but specific advice should be sought in situations where knowledge is to be created wholly or partly in foreign legal jurisdictions.

ABSOLUTE SECRECY

By keeping an idea or invention secret an inventor runs little risk that anyone else will see or find it and therefore be in a position to steal it. Such IP may be preserved until its owner chooses to divulge it. There is, however, little prospect for commercial exploitation of IP whilst protected in this way and, as secrets are difficult to protect for long, it runs the risk of misappropriation or loss. More seriously, the owner of such secret IP will find there is nothing they can do if another person independently develops the same idea and decides to make it public. Independent discovery or rediscovery of ideas is more common than generally realized. Researchers in the same industries working for competing organizations are often directing their efforts at the same problems – it is not surprising that they sometimes 'trip over' each other's inventions.

The point has validly been made that the mechanical protection of trade secrets is likely to be overcome after a period of time:

> The physical restraint upon unauthorised dissemination is, however, subject to one important weakness: it would currently seem that there is no electronic device or process so sophisticated that it cannot be neutralised, eroded or reversed by the exercise of sophistication equal to that of its original conception.[2]

There is a place for absolute secrecy in the field of contract research and other forms of knowledge-buying, but probably not where the technology/knowledge is of critical importance to the client organization. Absolute secrecy is easier to achieve for internal rather than external projects. In both cases, and especially in contract research, the buyer must assume that ordinary staff turnover and possible lack of loyalty to a previous employer (or a previous employer's client) means there is every likelihood of information leakage. Consider carefully how secrecy can best be preserved.

2 J. Philips and A. Firth, *Introduction to Intellectual Property Law*, 2nd edn, Butterworths, 1990.

COPYRIGHT

This is the monopoly use of one's personal creation (literary, music, art or software) in order to prevent others from exploiting the work produced by the IP owner, but without the right to prevent the exploitation of identical or similar work produced through the independent intellectual work of others. In theory this may seem a less adequate form of protection than absolute monopoly in the form of patents, but in practice it is rare for identical works to be created independently.

Copyright is, primarily, a right not to be copied without permission. It protects the form in which ideas are expressed, but not the ideas themselves. It gives legal rights to the creators of certain kinds of material (or their employers) so that they can control the various ways in which their material may be exploited. The rights broadly cover copying or reproducing, adapting, publishing, performing and broadcasting the material. It protects original literary, dramatic, musical and artistic works; sound recordings; films (including video) and broadcasts (including cable and satellite). Computer programs are protected in a similar way to literary works. Copyright protection is automatic in the UK, so no registration of copyright is required (indeed, no such register exists). In the UK copyright lasts for 50 years from the death of the author, except for computer programs where it lasts for 50 years from the end of the year in which the program was created.

To determine the extent of international copyright it is necessary to look at the membership of relevant multilateral IP conventions. It is prudent that all *published* material should bear the international copyright convention symbol ©. However, the © symbol should not be used outside the UK on material *that is not to be published.* Under the laws of many countries, the use of the © symbol implies publication and this could be undesirable in regard to plans or other information that is to be kept confidential. This is considered particularly important in regard to computer programs, where often a catch-all IPR clause similar to the clause below will be found:

> All intellectual property rights in the software and user documentation are owned by ... or its suppliers and are protected by United States and Canadian intellectual property laws (including patent, trademark and copyright laws), other applicable intellectual property laws, and international treaty provisions. ... retains all rights not expressly granted.

In situations where copyright material is to be used outside the UK it is prudent to place a prominent legend such as 'Copyright Reserved' on the material and back it up with an IPR clause similar to the one above.

It is prudent to declare your copyright by using the international copyright convention symbol © followed by your organization's name and year of first publication except as noted immediately above. You may need to give third-party contractors precise instructions about assignment of newly created copyright to your organization.

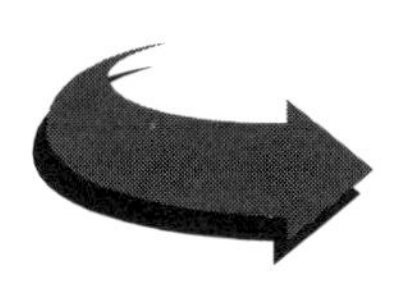

For third-party supply contracts that involve the creation of software, include in the contract a clause assigning copyright to your organization. It is also important to obtain a perpetual all-purpose license in any background software that may have been used by the contractor and incorporated in your software.

PATENT OF INVENTION

A patent is a grant by a government of monopoly rights for a period (in the UK) of up to 20 years in return for full disclosure of an invention. It prevents others, without permission of the patent owner, using the invention defined by the wording of the patent. To be patentable an invention must be new, not obvious, and contain an inventive step. It protects the concept of practical commercial utility rather than the appearance of a particular article (in the UK this can be protected by unregistered designs – see below). Detailed rules vary from country to country but a patent can protect various applications of technology, a method (whether chemical or mechanical) an apparatus or the use of an apparatus.

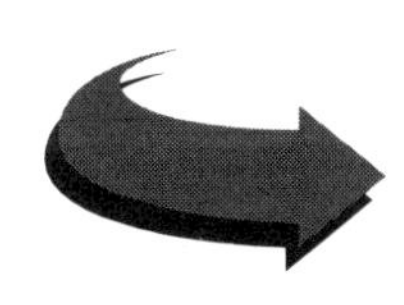

A contract under which knowledge/IPRs will be procured should clearly specify the supplier's responsibility either for filing a patent on the client's behalf, in the name of the client, or for cooperating with the client in the patent filing activity.

Ensure that the supplier is briefed as necessary – and that they brief their employees – as to their and their responsibilities for maintaining strict confidentiality. It is extremely important not to disclose any aspect of an invention outside your company (and by extension, the supplier's company) unless under a very strict confidentiality agreement with whomsoever, until 18 months have elapsed after a patent application has been filed. Earlier disclosure can cause a patent application to be rejected.

An invention by a supplier will be owned by that supplier unless the supply contract makes it clear that the client owns all such inventions.

RIGHTS OF EMPLOYED INVENTORS

In the UK an invention made by an employee belongs to the employer if:

- the invention was made in the course of the employee's normal duties or in the course of duties falling outside his or her normal duties but specifically assigned to the employee and the circumstances were such that an invention might reasonably be expected to result from the carrying out of his or her duties; or
- the invention was made in the course of the employee's duties and at the time of making the invention (because of the nature of the employee's duties and the particular responsibilities arising from those duties) he or she had a special obligation to further the interests of his or her employer.

It can be seen that ownership (and to some extent, compensation) depend on the employee's job description. If the job is defined narrowly, it is probable that an invention not covered by the duties detailed in the job description will belong to the employee. Conversely, if the job is scoped very widely, most activities of the employee will be covered by it and so any invention in these circumstances will belong, prima facie, to the employer. Note, however, that too wide a definition could well be overturned as a term that diminishes the employee's rights. Section 42 of the UK's 1977 Patents Act states that a contract (a) with the employer (alone or with another) or (b) with some other person at the employer's request or pursuant to the employee's employment contract which diminish the employee's rights is unenforceable.

If an invention that belongs to an employer is patented and the patent is, having due regard to the nature of the employer's business and to the business' size, of outstanding benefit to the employer, a British court may order the employer to compensate the employee who made the invention. (NB The position of an inventive employee who is acting under a contract between their employer and a client that assigns all inventions to the client is not clear if, for example, the invention turns out to be of outstanding benefit to the client.)

Contracts to be performed in foreign countries will be subject to the rules of that country relating to ownership of inventions, industrial designs and copyright etc. Even if a 'neutral' legal domain is agreed in the contract, it may in practice be difficult and expensive for the client to have the contract properly enforced.

It might be possible to seek an indemnity from the knowledge supplier to the effect that the client shall be fully and effectively indemnified by the supplier against any claims by the supplier's employees or former employees. However, this might sour pre-contract negotiations and should only be considered if there is a strong likelihoood of problems in this regard.

Caveat emptor! If the contract is to be performed by a foreign supplier, require the supplier's personnel department to comment on their employees' employment rights in relation to inventions. Then get your legal advisor to verify this, if necessary with a law firm in the foreign country. You may need to see a copy of key personnel contracts of employment. Again this could sour relations pre-contract *but may be necessary* – so be prepared for an argument. If you cannot get full agreement on this then an indemnity, as suggested above, might be a compromise solution.

TRADE SECRECY

The law of confidence or trade secrecy complements the patent system and may provide an alternative type of legal protection for inventions. It will prohibit, by injunction, the use or disclosure of information not in the public domain, if that would be a breach of confidence. The extent of information which may be covered is quite wide and includes the all-important concept of know-how, which comprises information that is secret, identifiable and substantial (that is, not trivial and *is* commercially useful) and information which gives its possessor an opportunity to obtain an advantage over competitors who do not know or use it.

The law will protect confidential information from publication or use by a party who receives it in confidence and undertakes (expressly or by implication) not to disclose or use it without the consent of the person who supplies the information.

The law of confidence is concerned as much with ideas as with their expression and will protect both of these to the extent that the necessary conditions for the protection of a trade secret are present. There is no property in information, as such, so it cannot be stolen.

Secret information should only be transferred under contract. Confidentiality agreements must be signed by a senior official (normally managerial grade) before such information is transferred unless other suitable contractual arrangements are already in force.

REGISTERED AND UNREGISTERED DESIGN RIGHTS

- *Registered Design Rights in Aesthetic Designs* – this is a protection for the outward appearance of an article and lasts for 25 years.

- *Unregistered Design Rights in Industrial Designs* – this is a uniquely British automatic right protecting three-dimensional industrial designs from copying 'for commercial purposes'. It lasts for 15 years from the date of design or 10 years from first marketing.

UTILITY MODELS

Some 50 countries offer specific protection for 'utility models' – these are often described as 'lesser inventions' and protect items that do not satisfy the inventive step requirement of the normal patent system. The current German system, which is adopted with variations by many other countries, protects tools and implements, chemical substances, drugs, electrical circuits, articles of daily use or parts thereof, foodstuffs etc. *provided that the subject matter of the application* (a) exhibits a new configuration (b) is capable of industrial application and (c) is based on an inventive step.

The initial period or protection is three years and this can be renewed for further periods up to eight years. Under German law the registered owner has the exclusive right to use the utility model and to prevent rivals from:

- producing, offering or marketing the product
- importing or stocking such product.

The utility model system may be an attractive short-term means of protection. In a world of ever shorter product life cycles, utility model protection of research results in those countries that offer it may be better value for money than a full patent.

FOREGROUND AND BACKGROUND IPRs – OWNERSHIP ISSUES

This is probably the single most difficult issue encountered by client organizations *where they are buying knowledge with a view to future commercial exploitation.* It is axiomatic that a supplier will be attractive to a client organization because of, and who will be persuaded to award a supply contract on account of, the supplier's existing knowledge of the subject covered by the contract. In principle, the greater the experience and skill level of the supplier (and therefore the greater the level of pre-existing knowledge it brings to the contract work) the higher premium the client may expect to pay.

From the supplier's point of view, it has a legitimate interest in:

- preserving and extending their knowledge base
- not being used/abused as a 'knowledge mine' by clients
- not releasing valuable know-how to clients without suitable recompense.

Whilst recognizing these legitimate interests, the client also has commercial and technical interests, which are equally legitimate:

- it is funding the extension/deepening of the supplier's knowledge base
- what emerges from the contract must be of use to the client without the need to pay additional sums for 'corollary' IPR
- it is often true that the client brings their own background know-how to the

research project which adds value to it and which directly or indirectly benefits the researcher/supplier.

WHAT ARE FOREGROUND AND BACKGROUND IPRs?

The term often used to describe IP that existed prior to the commencement of a supply contract (typically a research or consultancy contract) is *pre-existing rights.* The term acknowledges by implication that from the supply contract new IP rights will come into existence which may, to a greater or lesser extent, be dependent upon work carried out before which happens to be the IPR of *somebody.*

In October 1988, the Commission of the European Communities (CEC) published a useful *Model Contract* for use in EU R&D projects which, apart from being a useful benchmark of all R&D contracts (especially collaborative ones) provides a helpful definition of 'background' and 'foreground' IPRs. These definitions are as follows:

- *Foreground information* – means information including all kinds of results, generated by any contractor or third parties working for any contractor, in the execution of [this] contract.
- *Background information* – means information, excluding foreground information, owned or controlled by any contractor in the same or related fields to the research under [this] contract.
- *Foreground patents* – means patent applications, patents, copyrights, plant variety rights, and other similar statutory rights for inventions or improvements made or conceived by any contractor or any person engaged or employed by any contractor in the execution of [this contract].
- *Background patents* – means patent applications, patents, copyrights, plant variety rights and other equivalent statutory rights excluding foreground patents, owned or controlled by any contractor in the same or related fields to the research executed under [this contract].

The CEC model contract goes on to set out how background and foreground rights are to be used by and between the various collaborating partners and establishes a right of all parties to a royalty-free licence to use all foreground information (that is, information which emerges from the project) to the extent necessary for such other parties to carry out their work. Regarding background information/patents, the contract sets out that each of the contractors shall, to the extent they are legally allowed to, upon request and on non-discriminatory transfer conditions, make available background information *and* grant non-exclusive licences to its own background patents to the other contractors in the project to the extent necessary to enable them to carry out their work under the project. Foreground information and foreground patents are owned by the party generating them under the project. Background information and patents continue to be owned by the party to whom they originally belonged, subject only to the necessity to grant sub-licences as described above.

A VEXED QUESTION – OWNERSHIP AND USE OF BACKGROUND RIGHTS

The first thing a client organization must do is to establish the extent to which any IPRs that emerge from his contract are, or may be, fettered by the supplier's pre-existing IP rights. Dependent on those rights, their nature, their extent, their value and the negotiating strengths of the parties, the client must then secure a royalty-free licence to use those background rights in order to *exploit the full value of the IP that emerges from the contract.*

But how do you establish what those IPRs are? How do you establish their extent? This again is a difficult issue. The best answer is a two-pronged attack on the problem:

1. Persuade the supplier to discuss and identify the principal IPRs that it will use in carrying out the project. Decide on an appropriate mechanism for enabling the client to utilize those IPRs necessary to fully exploit the 'foreground' IPRs (or 'results'). Identify those IPRs in the contract document.
2. Incorporate into the contract document a requirement that subject to any pre-existing rights vested in the supplier, all IPRs arising from the contract shall be the absolute property of the client and list in the contract those background rights that are specifically excluded.

The supplier may complain that it is not possible to list all the IPRs that are applicable: they may be far too extensive to list. What happens if they inadvertently omit one? Some may be too confidential to discuss openly, and so on. The client may have some sympathy with these difficulties, but they are not insuperable. Where there's a will, there's a way...!

This can be strengthened with a contractual clause along the lines:

> The supplier shall notify the buyer in writing as soon as possible after the commencement of the contract, or as soon as possible after (during the course of the contract) the supplier becomes aware of any pre-

> existing IPRs vested in him which are applicable or may be applicable to the work carried out under the contract. All property, right, title and interest in any IPRs not so notified is hereby irrevocably assigned to and shall vest in the client. The supplier shall carry out such further acts and execute such further documents as the buyer may reasonably request him to do to ensure the proper terms of and to effect the purpose of this clause.

NB A clause similar to this is featured in the model contract which forms part of the Toolkit in Gower's *The Outsourcing R&D Toolkit* – see details in Appendix 1.

This may prove controversial but is not unreasonable, bearing in mind the quite legitimate need of the client to protect against any possibility that the investment it has made may be lost (or forced to be increased) by a supplier that seeks to deny the client necessary access to pre-existing rights.

A typical example of the problems that can be experienced over the use of pre-existing rights is in the field of computer software, where a supplier may use pre-existing software as a basis for work to be done for a client. The question then emerges, what happens if the client wishes to rewrite or amend the software, or to get others to do so? They may be prevented by the supplier on the basis that it will not allow source code to be used without further recompense.

Another example is where the use of the results in some way depends on a written document or a piece of hardware which the client did not, and could not reasonably have, known about. When the client tries to exploit the results for their own purposes, they may find an unexpected cost payable to the supplier for access to this pre-existing IPR. In extremis, the supplier may deliberately exploit this power over the client.

The answer to this vexed question is to ensure that sufficient ownership, or at least an appropriate royalty-free perpetual licence, vests in the client at the conclusion of the contract.

CHAPTER 4

Intellectual Property Rights: Current Developments

POSSIBLE CHANGES IN THE US PATENT REGIME

Thomas Jefferson, drafter of the American Declaration of Independence, examined his nation's first patent in 1790. Today the US has patented some 6.5 million inventions. At any one point in time there are currently some seven million applications underway, reflecting today's knowledge explosion. Proposals aimed at liberalizing US patent law have proved controversial: US law presently allows an inventor to keep a patent application secret for as long as it takes the Patent and Trade Mark Office to issue or deny a patent. This period is supposed to average two years but in practice it averages four years and, for a very complex technology, five to ten years is common. (The first patent on laser technology took 20 years to process.) Until 1995 a US patent was effective for 17 years from issue but that was amended to 20 years from filing of patent application, following a US/Japan trade deal. Current proposals on patent reform are aimed at:

- making the Patent and Trade Mark Office a quasi-private corporation
- widening the basis on which patents may be challenged
- requiring patent applications to be published 18 months after filing.

Quasi-privatization is aimed at speeding the patent process by reducing bureaucracy, but critics claim that a less deliberative patent grant will lead to less effective patent coverage and may also favour big corporations over smaller ones or individual inventors. Why? They will have more clout with the private patent agency because, of necessity, they will be its main paymasters.

The second proposal (widening challenge bases) is aimed at quicker and better diffusion of technology – particularly important where product life cycles are shortening. Opponents claim, however, that this will scare off individual inventors and small corporations/venture capitalists by making the process less certain and thereby damaging inventiveness.

The third proposal (early publication) is aimed at reducing 'submarine patents' from surfacing years after a technology has been widely adopted, thus allowing patent owners to extract potentially ruinous royalties from others who have innocently been using the technology in the interim. But this again has been criticized: as one US commentator, Dana Rohrabacher, a Republican from California and one-time speech writer for Ronald Reagan has said, likening the proposals to a David (small inventors) versus Goliath (big corporations) battle, the proposed changes will 'gut our patent system and put our economy in jeopardy' because 'every international thief and gangster corporation is just waiting … for us to disclose all our information before patents are issued'.

The Twenty-First Century Patent Coalition, a group of large corporations that support reform proposals, claim that reform will help US inventors bring ideas to market more quickly with less uncertainty over patent challenges. Industry needs ever more inventions to survive in the global market: IBM, for example, earns more than 50% of its revenues from products that have been marketed for less than 12 months. It seems there is a majority of large US corporations in favour of change.

Of interest to organizations involved in technological or scientific research and development, such organizations should monitor developments in the US and other patent regimes. Shortening of patent application secrecy may mean that the US becomes a less attractive proposition as a venue for contract research; for example on a long-term contract (three years or more) the results may be published before the contract is complete! It may be necessary, where there are material differences between the patent regimes of various trading blocks which render one more beneficial than another for long-term research, to shift R&D facilities to the more beneficial regimes.

PROBLEMS WITH EXISTING PATENT ARRANGEMENTS

Intellectual property rights are the foundation of the information revolution and the technological edifice it has built. Skills and knowledge have become, arguably, the *only* source of sustainable long-term competitive advantage and therefore knowledge is at the centre of the modern organization's success – or failure. What has changed from the commercial and industrial world that has dominated in the last 100 years?

- raw materials can be traded globally and their long-term price trends are downwards
- capital is now a global commodity
- truly unique technical equipment which once provided competitive advantage is now virtually unknown.

If materials, capital and productive capacity are now unlikely to be the primary foundation of economic success, then they have been replaced by knowledge and knowledge workers. Microsoft owns little of value except knowledge, and its founder, Bill Gates, the archetypal knowledge worker, is now one of the world's richest men. The industrially developed nations' fastest growth industries are:

- microelectronics
- biotechnology
- energy conversion
- speciality materials
- telecommunications.

All of these are knowledge-based industries. If their IPRs can be easily stolen and copied then fortunes can be made and lost. More importantly, the economies that depend on them may be irreversibly damaged.

IP is increasingly important as a source of licence revenues. For US electronics giant Texas Instruments the income derived from aggressive licensing has in some years exceeded that from operations. Other corporations are expected to increase their efforts to derive income from licensing. IP is increasingly important in corporate strategic planning; companies like Intel have large legal budgets to defend their own IPRs, and have been accused of harrying competitors on IPRs with a strategy of creating uncertainty, higher start-up costs and slower market penetration for rivals. The world's most successful companies today are those with a lock on some form of knowledge.

RESTRICTIONS IN THE FLOW OF KNOWLEDGE

From 1945 until the mid-1980s technological knowledge flowed in a comparatively unrestricted way around the world, especially US knowledge. The American government funded most basic research in the US and, with the exception of military technology, encouraged worldwide dissemination. During the Cold War era, economic success of friendly countries was considered important to the US's own security interests. There was, in any case, a comfortable assumption pervading the US that since the rest of the world would be permanently behind the US's technological lead, unrestricted dissemination of today's IP to friendly nations was a sound policy. During the era of the 'brain drain' to the US, America's belief in the unassailability of its own technological superiority was probably well founded. At the end of the twentieth century, however, the US government was cutting its investment in R&D. Federal R&D investment fell a significant 14% by 2002 compared to 1997 levels. It may confidently be predicted that less *new knowledge* will be available in the public domain – and what there is will be more restricted.

As the US government slows R&D spending, there is a belief that US corporates will fill the void, though not necessarily in basic science. There is, however, a very real danger that, without stronger and more focused systems of IP protection, companies will opt for more secrecy as a means of protecting, and maximizing returns on, their R&D investment. It is well known that a rival's research programme that can identify what is already known by its main competitors can often leverage that knowledge. Without access to others' knowledge, a research programme may be forced to 'reinvent the wheel', thus slowing new product development and increasing costs. It has been estimated that in the US, 73% of private patents are based on knowledge obtained free of charge from public (mainly governmental) sources. This statistic alone suggests that widespread secrecy about new knowledge will hamper the development of the next generation of new knowledge.

JUST WHAT CAN AND CANNOT BE PATENTED?

The development of new plant varieties, the isolation of genetic coding, animal cloning and the ability to download complete libraries of information via the Internet all challenge the traditional concept of IP protection. What can and what cannot be patented? And what does copyright really mean in the age of the 'world wide web'?

It has recently been argued (for example by Lester C. Thurow in 'Needed: A new system of intellectual property rights', *Harvard Business Review*, October 1997) that some differentiation is required between fundamental advances in knowledge and logical extensions to existing knowledge. Each of these deserves, he suggests, a different kind of patent. Thurow cites the case of a doctor of medicine who noted a relationship between a particular hormone and a congenital birth defect. He was granted a patent for this observation, although by itself his test had too many false positives to be useful. Later development demonstrated that if his test was used along with a brace of complementary tests, they would predict whether a baby would be born suffering from Down syndrome. The doctor is currently suing to obtain a $9 fee from every laboratory that uses 'his part' of the successful test. If he wins, Thurow notes, the cost of testing will more than double. Whilst acknowledging that the physician probably did deserve IP rights over his observation, Thurow suggests:

> They should not be the same kind of rights as those granted to someone who invents a new gene to replace the defective one. Noting what an existing gene does is simply not equivalent to inventing a new gene. Such distinctions are necessary, yet our existing patenting system has no basis for making them. All patents are identical – you either get one or you don't.[1]

Knowledge buyers need to be clear as to whether the information they seek via a third-party knowledge provider, such as a *contract research organization,* is likely to result in a truly inventive step (and therefore likely to be patentable) or will cast new light on existing knowledge, in which case some lesser form of confidentiality may be a better way to manage the information.

Copyright may now be an inappropriate form of protection for data that can be scanned or downloaded, manipulated, repackaged and resold. The barrier to copying what was traditionally printed media may no longer be adequate.

Where knowledge buyers are relying on a written report and/or data to satisfy their specification, they should determine whether all or any of the content should or could be protected other than by copyright. For example some phone companies, which invest large sums of money building databases of phone numbers in telephone directories, are now inserting bogus numbers in order to prove in court that competitors have *not* generated their own list of names and numbers.

1 Lester C. Thurow, 'Needed: A New System of Intellectual Property Rights', *Harvard Business Review,* September–October 1997, p. 96.

THE BATTLE TO JOIN THE 'DEVELOPED WORLD'

The acquisition of knowledge is essential for industrially developing countries and undeveloped countries. At the beginning of the twenty-first century, a number of knowledge acquisition strategies are prevalent:

- *Knowledge as a 'fee'* for market access rights: nations acting as monopsonists (buyer controls the market) provide opportunities for investors/sellers in their ostensibly lucrative market. But before contracts are concluded, technology transfer is negotiated. For example in 1997 between Boeing and the Chinese government, where access to the Chinese aerospace market was granted only with significant technology transfer. Part of Boeing's strategy for allowing this was to pre-empt Europe's Airbus Industrie from doing the same thing. To operate in these markets, sellers must swap knowledge for access rights.
- *Countertrade requirements*: a more sophisticated version of the above, typically found in large-scale infrastructure or military projects, where a given percentage of a particular project's value must be spent by the seller in the host country. Know-how transfers are normally backed onto countertrade deals and are closely monitored to ensure terms are met by the seller.
- *Copying*: most rapidly developing countries are heavily involved in direct or indirect copying, either legally or illegally. In many of the cultures in which knowledge owners must operate, there is less regard for personal property rights – and especially intellectual property rights. This is a cultural issue and needs to be recognized as such.
- *Universities*: the establishment of, or increased investment in, institutes of higher education by industrially developing countries is often coupled with sending an elite cadre to study in 'first world' universities. In 1997 there were calls in the US Congress to debar foreign students from US universities in order to prevent a perceived threat of taxpayer-funded R&D and advanced technology know-how 'leaking' abroad.

Knowledge buyers should ascertain whether potential third-party knowledge vendors such as contract research organizations may be a particularly potent source of knowledge leakage and what practical 'Chinese walls' may be erected to safeguard new knowledge.

All these developments, under the general banner of 'globalization', are putting strain on the patenting system. Patent costs tend to be high, yet offer less protection than in the past. IP protection systems will not work unless most governments agree to implement them. A law that does not exist, or is not enforced, in country A is, in practical effect, a law that cannot be enforced by aggrieved country B and there is a danger of a shift of production to countries which less rigorously enforce IP laws. In extremes this may lead to trade wars, which is to the long-term detriment of IP owners.

SOME SUGGESTIONS ON IP LEGAL DEVELOPMENT

It is no surprise that the debate on the development of IP rights is led by the US, the country that has the most to lose from the globalization of knowledge. The following arguments are prevalent in the US:

- To develop new products/processes/knowledge, individuals or corporations must have financial incentive to undertake the costs, risks and effort in R&D.
- As government funding of R&D decreases, the need for stronger incentives for private sector investment in R&D grows; the basic incentive has traditionally been monopoly rights leading to a maximum return on investment. A corollary of reduced state investment in R&D (less 'free' knowledge) is stronger private monopoly rights.
- In tension with the above, once IP exists the social imperative is to encourage use and rapid dissemination to the benefit of society at large. (If a real cure for cancer was discovered, it is unlikely that the state would allow monopoly rights to exist on the invention.)

  New IP rights legislation must balance conflicting objectives – stronger monopoly rights versus the social imperative. Legal thought processes alone are unlikely to lead to the optimum method of striking this balance. Lawyers tend not to think in economic and technological terms – their imperative is to establish new concepts within existing legal frameworks. A better approach might be, via state legislatures, to weigh the underlying economic realities of an industry against the national interest (or in an EU context, the European interest) to determine what IP incentives are necessary for successful technological development.
- Private monopoly power may be deemed to be less threatening to society at large now that anti-trust laws are better developed, governments are more willing to intervene and the free media is willing to focus on abuses of power. Technological proliferation has, in any case, undermined monopoly power by enabling 'technology substitution' (akin to the economic concept of 'product substitution') to a certain extent.

- IP laws need to be enforceable, or they should not be laws – so runs the slightly simplistic argument. But this still leaves one question unanswered – How do you protect what ought to be a *right* in a situation where enforcement is not possible? A partial answer might be by intergovernmental agreement, and, perhaps, by the levying of punitive international taxes against corporations that flout whatever agreements exist. This, however, will require a stable international situation and an unprecedented meeting of minds between governments on this complex issue!
- IP dispute resolution needs to be quick and efficient – a suggestion has been made by academe that patent fees should be set higher to ensure a higher quality (quicker and more certain) service. Variable fees, akin to a progressive tax system, could equalize burdens between large and small companies, as well as individual inventors. IP owners of inventions with short economic lives would benefit from speedy dispute resolution, so the argument runs. However, it is uncertain that the business community will welcome the idea of higher patent fees.
- Certain classes of information should be public domain: it is argued that the basic sciences should be publicly funded whilst developed products should be subject to private monopoly rights. This seems to be a sophisticated plea for a return to the classic pattern of science – universities and public bodies do research; industry undertakes development and brings products to market. The first objection is that this runs counter to the fundamental economic drift in most countries (especially the former G7 countries) that there should be a general withdrawal of government from economic activity. The second objection is that the suggestion requires agreed principles to determine what should be publicly available and what should be private – together with a methodology to reward those whose inventions must be made public domain. Then there is the question: how do countries decide what basic research is to be undertaken?

These suggestions would require increased public funding, compulsory purchase of certain classes of information and some form of adjudication to safeguard the interests of inventors, none of which seems particularly likely in the business/political environment of the early twenty-first century.

- A new global system of IPRs has been argued for – reflecting the needs of developed *and* developing countries. The developing nations' need for low-cost/high-quality medical technology/products is of an entirely different magnitude to their need for low-cost TVs. An IP system that treats these needs as equal fails both developing and developed countries and the suggestion has again been made that the relative importance of a technology to a recipient country should attract variable levels of royalty payment.

Whilst this concept has theoretical merit, industrially developing countries may view this as being in some sense 'imperialistic' or paternalistic on the part of developed nations. It also runs the risk, if it *increases* the net financial outflows for technology of, in effect, increasing the already crippling burden of Third World debt. Why should developing countries opt to pay more when technology piracy is already a viable option?

- Multilateral patents could be devised to tailor IP protection to:
  - different industries
  - different types of knowledge
  - different types of invention
  - different parts of the world.

A cursory comparison of the needs of the pharmaceutical industry with the needs of the electronics industry quickly reveals that electronics need fast patent grant, fast dispute resolution, short duration patents and (perhaps) lower-cost patents – because revenue earning and product life cycles are short. Pharmaceuticals, by contrast, require long-term protection, may be more tolerant of patent and dispute delay (because clinical trials of necessity span many years) but require very long protection and (perhaps) new mechanisms to enable government agencies to grant licences of right to other manufacturers whilst also providing an agreed system to pay royalties to the prime patent owner.

Similarly, new inventions could be granted a different type of patent to logical developments of existing knowledge which represent a lower level of inventiveness. The argument is that true new inventions obtain a traditional long-term patent, because the inventor has arguably encountered most risk and uncertainty in undertaking the inventive step, whereas a logical extension to existing knowledge is rewarded with more limited monopoly rights for the owner – shorter term but probably cheaper protection.

These are interesting and useful contributions to a growing debate but there are no easy answers to the increasingly urgent need for change in IPR protection across the globe. Knowledge buyers should be aware of these debates and monitor them; their outcome will surely affect the long-term knowledge-buying strategies of organizations engaged in buying-in research and development services at the cutting edge of knowledge.

IP AND TAX HAVENS

In January 2004 talks between global pharmaceutical giant GlaxoSmithKline and the US Inland Revenue Service (IRS) over a long-running tax dispute broke down. GSK's share price dropped by 2% over the threat that the IRS might demand a back payment of a cool $5 billion in 'unpaid' taxes, penalties and interest. The dispute concerned the somewhat arcane area of transfer pricing, and particularly, the rate at which GlaxoSmithKline charged for marketing services supplied by its US affiliates from 1989 to 1996. The IRS case was that the GlaxoSmithKline transfer pricing rate was too low, which in turn greatly understated the company's income and so the tax paid. At the time of preparing this book the case remains unresolved.

For multinational corporations transfer pricing is the most important current tax issue. According to industry experts, the IRS's decision to take GlaxoSmithKline to court reflects new thinking by the US tax authorities. The rules would radically change how the US tax authority treats services supplied to parent companies by affiliates in other tax jurisdictions, including jurisdictions that offer sufficient tax advantages to be considered as tax havens. At the time of writing there is concern that the rules may not be workable and that they might subject multinationals to huge US tax increases.

Under present EU and US tax rules, transfer pricing of services can be reported at cost providing those services are considered to be integral to the business. Those that are integral must be priced as if they were offered by a third party, which usually amounts to cost plus a mark-up that would be appropriate in arm's length transactions. What is integral is, of course, subject to interpretation, as is the question of what is a suitable mark-up.

The proposed new IRS rules eliminate the safe harbour for non-integral services and instead require these services to be priced at cost plus. The mark up must be demonstrably close to what is available from a third party. If a third party mark-up exceeds 10% a company would have to use one of four other, rather more complex, pricing methods:

1. manufacturing, production, construction or extractive pricing;
2. reselling, distribution, sales agency, purchase agency or commission arrangements;
3. R&D, engineering or scientific pricing;
4. insurance, reinsurance, or financial transactions including guarantees.

The IRS believes that additional tax burdens for multinationals will be realistic and appropriate. Looking at the GlaxoSmithKline case again, observers note that the company relies extensively on marketing and distribution in the US market through local GlaxoSmithKline affiliates. However, the company's R&D activity is conducted mainly in the UK. It is the IRS's contention that much more value for GlaxoSmithKline's drugs is derived from marketing (US) operations than from R&D (UK) operations. GlaxoSmithKline's transfer pricing does not reflect, according to the IRS, the value of work done in the US. Clearly if the IRS view is upheld, there could be added tax burdens for many multinational corporations. Under the new rules multinationals would have to add a mark-up based on an estimate of what it would pay a third party for the services provided by its local affiliates. But the company would have to prove that the mark-up was appropriate, based on market prices or 'comparables'. If it cannot do this with some degree of accuracy, additional tax may be due. Ignoring the mathematics, the sums at sake could be considerable.

The IRS is trying to overcome a legal precedent established in a *contract research* case in 1992. The case involved Westreco, a subsidiary of the global food giant Nestlé. At dispute was some contract research that Westreco undertook for its parent Nestlé, which was later reported in the annual accounts at cost plus mark-up. The IRS took the view that the mark-up was too low, but the court pronounced in favour of Nestlé. The IRS's real target in its current moves, some observers suggest, are US companies that have established manufacturing operations in tax havens so as to avoid US tax. Under the IRS's current approach, these companies could face hefty new tax bills based on profits generated in the US, if they are deemed to have transferred engineering or scientific know-how in the process. The IRS might then require a parent company to report not the cost of employee salaries who transferred the engineering or scientific know-how, but a proportion of the profit that would reflect the value of a typical licence agreement that a third party would pay for the 'intangible asset' involved. All very complex and somewhat arcane, but a subject worth noting for those who are involved in buying in knowledge services from overseas. A further danger is that if different tax authorities take different views on transfer pricing, multinationals could face double taxation.

Some multinationals have engaged in tax avoidance by compensating foreign affiliates for services at prices higher than appropriate.

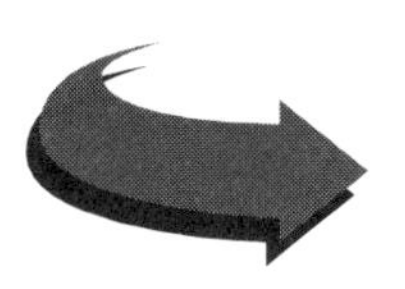

One abuse is to contract with an R&D affiliate in a high tax country on a cost plus basis, and later to restructure the terms once the R&D has proved successful to provide a subsidiary in a tax haven with much of the resulting profits.

Ensure that smooth tax consultants do not lead you to place knowledge-based work in offshore havens if there is any realistic prospect of domestic tax authorities taking a contrary view. You will know the real reasons why you shift work to different companies. Always remember that the long-term costs may outweigh short-term gains. Are you in this business for the long or the short term?

To conclude this section on the tax havens and the smoothness of some tax consultants, consider this quote from the UK *Financial Times* (special report on intellectual property, 30 April 2003). Quoting a tax advisor, the *FT* noted:

> By basing IP assets in a separate offshore entity companies can take advantage of low tax regimes by receiving income generated in higher tax areas. By having an income stream in a separate entity, you can value assets more easily and as a result use it as security in financing. And by having the assets in a low tax jurisdiction, it increases their worth in the event of a sale or takeover ... If a company such as Microsoft had developed its IP for its software products using a Bermuda IP holding company, and then received its license income through that entity, how much would the company be worth today?

THE EUROPEAN PATENT

On 3 March 2003 EU member states concluded 30 years of deliberation to reach an outline agreement on a single harmonized patent that will be valid throughout the EU. Companies in particular should benefit from a cheaper, more efficient one-stop shop for patent services, rather than having to apply for many patents throughout the EU zone. However, the patent will not come into force until 2006 at the earliest and the proposed European Patent Court will not be formed until 2010. There will be a need for transitional arrangements, yet to be defined.

Under new rules, EU companies will have to pay €25 000 for the new community patent. This is still higher than in the US and Japan, but is approximately half the average current cost of applying for patents across Europe. EU member governments hope this will encourage EU firms to increase R&D spending and help to convert the EU zone into the world's most dynamic economy. Time will tell whether this ambitious target will be met, but there remains a glaring disparity between the EU and the US in intellectual property. Although the EU and the US generate a similar number of patents, there are significant disparities between EU member states. For example,

Sweden generates twice the number of patents per head than does the US, but Portugal and Greece generate just 1% of the patents of their Swedish EU partners.

The difficulties of doing a final deal are not to be underestimated. There are cultural and philosophical differences between the approaches of the member states. In the past countries such as France and the Netherlands have looked favourably on claims of patentees and trademark owners. In the UK the interpretation has been less liberal. Furthermore, some EU lawyers believe that the benefits of a community patent are already beginning to be negotiated away. Community patents will need to be translated (expensively) into all the official EU languages (19 now that the EU has reached 25 member states).

For EU companies, the cost of safeguarding IP rights is just one of a growing list of bills that they must meet as 'knowledge-based assets' increase in significance. Spurred by the idea of 'buried IP treasures' larger businesses (and some smaller ones) have trawled their patent portfolios in the hope of rediscovering winning ideas that will have future commercial potential. Elsewhere in the EU discussion continues on possible methods for balance sheet valuation of IP assets, but there is as yet little consensus on the subject. Finally the whole question of IPR enforcement is under scrutiny. Attempts continue to be made at EU level to devise a viable centralized model. According to a Danish government study the annual economic gains could be as high as €21 billion across Europe. The accuracy of this has been questioned, but what *is* clear is that the potential economic benefits are very large.

CHAPTER 5

Why Buy Knowledge?

THE ROLE OF THE MODERN MANAGER

Managers are not employed to make the inevitable happen. Many business and organizational functions and departments carry out their day-to-day duties without much need for detailed management oversight. Managers are employed to steer the organization in ways that maximize the value represented by their organization's various constituent parts. The higher that a manager rises within an organization, the greater the responsibility for 'steering' and progressively less for day-to-day tasks. At whatever level a manager sits, however, there is a constant requirement to manage knowledge – to be aware of what knowledge matters now and what knowledge will matter in the foreseeable future. As the old adage has it: 'The fool learns from his mistakes. The wise man learns from other people's.' The role of the manager, at this level, is to anticipate knowledge needs for his or her organization.

Knowledge management activities may occasionally involve exploring new knowledge fields that, after evaluation, turn out to be irrelevant to the organization's objectives. The manager cannot be right all the time, but needs to develop the ability to review information and data from many sources to quickly establish those that are of lesser value, and those that will significantly impact organizational objectives in the medium and long term. Leveraging the corporate memory, the knowledge of staff and external sources of knowledge are all part and parcel of the manager's task.

The different potential sources of information for organizations are suggested in Figure 4. Managers must oversee and organize the processes of evaluation and analysis that convert information into useable knowledge. And knowledge must be lodged in the 'corporate memory' so it will be accessible to the whole organization.

KEEPING ABREAST

In Chapter 1 we reviewed the pace of change. It has rightly been said that 'change is the only constant'. But why do organizations need to buy in knowledge from external sources? Is it a sign of some deficiency that we are obliged, or feel obliged, to buy in information from third parties? Should our own organization not be able to supply all the information we need?

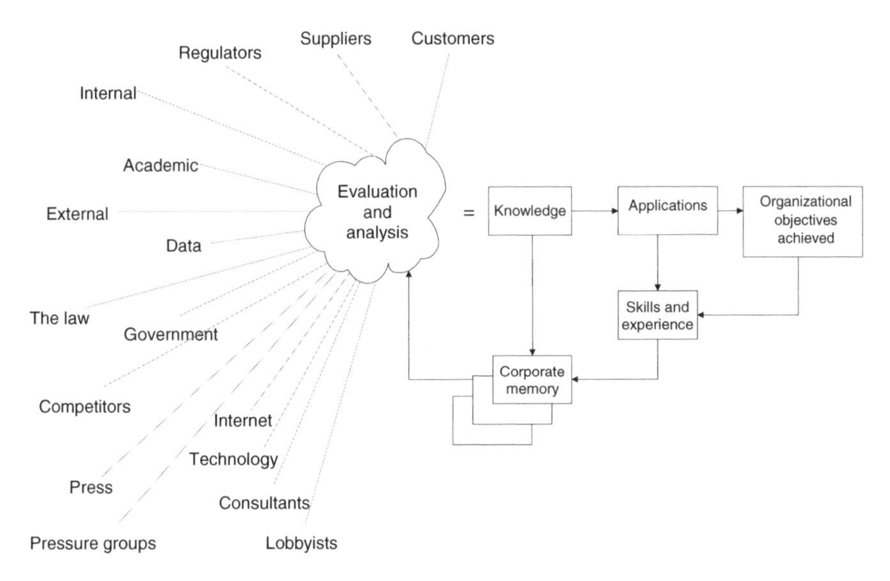

Figure 4 Organizations at work – from information to knowledge

A major study in 1992[1] identified technology acquisition, including *external* technology acquisition, as the most prominent issue in technology management. Table 7 is derived from this study and shows the five top technology management issues in large 'mul-tech'[2] companies (or MTCs) in Japan, Sweden and the US. It indicates that of the top five issues cited, technology acquisition was the most

Table 7 The most prominent technology management issues in multitechnology corporations (MTCs)

	Japanese MTCs	*Swedish MTCs*	*United States MTCs*
1	Diversification	Technology acquisition	Technology acquisition
2	Technology acquisition	Internationalizing of R&D	Organization/funding of R&D
3	Internationalizing of R&D	R&D productivity	Government policy
4	Investment in basic research	Shorter lead times to product development	Quality/availability of R&D labour force
5	Internal ventures	Quality control	Shorter lead times to product development

1 'External technology acquisition in large multi-technology corporations', O. Granstrand, E. Bohlin, C. Oskarsson and N. Sjoberg, Department of Industrial Management and Economics, Chalmers University of Technology, Gotenborg, Sweden, *R&D Management*, 22, 2, 1992.

2 The term 'mul-tech' is coined by the Granstrand study to describe corporations whose products and activities are based on several technologies – not necessarily 'high' technology – which these corporations bring together and synthesize. For these corporations the dominant characteristic – and the main challenge – is the need to diversify their technology base by acquiring increasingly complex and costly technologies.

prominent, being the primary responsibility for technology managers and, by extension, many other managers as well.

Whilst the Granstrand study referred specifically to technology acquisition, its lessons apply equally to externally sourced knowledge services. Buying in knowledge services is aimed, to a greater or lesser extent, at improving the technology base of the commissioning organization. The technology base consists of an organization's overall technological competence – its knowledge, skills and technology assets. As an asset of the organization the technology base can be acquired, developed and exploited in a variety of ways – always with the basic assumption that the organization's objective is to build up or maintain its technology base. There may be situations where an organization is prepared to see an erosion of its technology base. If so, this is likely to be in situations where market conditions are unfavourable over the longer term, leading to a strategic decision to undermine (divest or asset strip as appropriate) the technology base. The generic strategies for the acquisition and exploitation of technology are suggested in Table 8.

Table 8 Generic strategies for the acquisition and exploitation of knowledge

Knowledge Acquisition (Sourcing) Strategies	*Knowledge Exploitation Strategies*
Internal R&D	Internal exploitation (for example direct investment in production and marketing)
Hiring personnel with specific knowledge, skills, experience or insights	Creation of new commercial ventures (for example new firms)
Acquisition of knowledge-rich firms	Joint ventures
Joint ventures	Technology selling (performing contract R&D, licensing out, and so on)
Knowledge purchasing (contract R&D, licensing, and so on)	Divestment
Knowledge scanning	Storage, loss and leakage

The lessons of Granstrand have direct analogies for organizations that buy in knowledge. Within the various technology acquisition/exploitation strategies shown in Table 8 are varying levels of 'integration' between the corporation/organization, on the one hand, and its external technology environment on the other. 'Integration' is defined as the *closeness* of formal or contractual links between the various players in the technology transfer business. Table 8 suggests that within the *knowledge acquisition* strategies available, internal R&D represents the highest level of integration and knowledge scanning, the lowest. So with *knowledge exploitation* strategies, internal exploitation represents the highest integration and storage, loss/leakage, the lowest.

In knowledge acquisition a high level of integration is provided, for example, by internal R&D, where researchers are our employees, bound to us by a contract of employment and over whom we have a high degree of control. A low level of integration characterizes knowledge scanning, which is simply a systematic review of available sources of information about a subject. There is no contractual link between our organization and the information being scanned.

In *knowledge exploitation* there is a high degree of integration where we exploit the knowledge internally – but integration is lost where we allow the knowledge to leak away! Note that storage, loss and leakage is not a strategy for exploitation but represents the real world where information inevitably leaks after a period of time, for example through the knowledge-scanning efforts of others or through normal staff turnover.

The level of integration found in *joint ventures* cannot easily be characterized. There are varying types of joint venture in which knowledge may be acquired and exploited and they may not always be formalized in a contractual sense, for example joint ventures with competitors, with suppliers and with customers.

Integration, then, is the *degree, extent and closeness of the connection between players in the knowledge transfer field* and is an issue where new knowledge/ intellectual property is being created. When a contract is let, two parties become integrated, albeit temporarily, by the exchange of promises. It is not possible to cover every contingency in a contract, so an element of trust must exist between the parties. The nature of a commercial contract is to express a common will and intention between buyer and seller. Different types of contract reflect differing levels of integration. Thus, very specific and inflexible contracts may be appropriate for arm's length transactions (a minimum level of integration) while general, flexible contracts rely on a considerable measure of trust, perhaps suggesting a greater level of shared will and hence, greater integration.

Integration is based, to a large extent, on assumptions about shared will and intention. In the employment contract, integration assumes an employee's loyalty to his or her employer. Loyalty to the employer is important in the context of creation of commercially valuable intellectual property. The courts increasingly use the employment relationship of parties to assess whether technical information has been used or disseminated in harmful ways, especially where such information has found its way into a competitor's hands.

The contract of employment has been argued as *the core contract* which gives reality to the concept of a firm as a 'living' entity. Certainly it is the closest form of integration in a commercial contract. Other forms of integration exist, as suggested above – networks of buyers and sellers, trade and professional organizations, research 'clubs', joint ventures and so on. Some of these relationships become complex and long standing, so much so that it can become difficult to establish where the boundary of the firm lies. Where these relationships exist, the contract of employment assumes even greater importance, binding an employee to act in the interests of his employer and not to be disloyal in the event of a dispute between his employer and some third party. Such conflicts can and do arise where complex levels of integration exist.

Understanding the level of integration in a given situation, rather than assuming it or ignoring it as an issue, helps the knowledge buyer to formulate optimum strategies to acquire and leverage knowledge. Integration is a minor issue, but should not be overlooked.

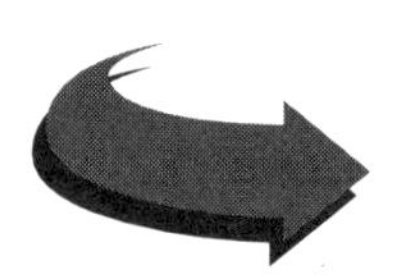

Knowledge buyers should evaluate the levels of integration already existing in their particular context. Where a higher level of integration appears necessary, consider the options available to achieve this.

KNOWLEDGE ACQUISITION STRATEGIES AND TYPES OF INTEGRATION POSSIBLE

Different knowledge acquisition strategies, as already noted, involve differing levels of contractual integration. Thus from highest to lowest integration, the strategies available are:

- *Internal research* – Based on an employment contract with a high degree of integration. Although the relationship is 'authoritarian', a strong element of trust is expected.
- *Acquisition of knowledge-rich firms* – An extension of the firm, since the acquired firm also uses employment contracts. Note that the ownership may not be 100% and it will take time for employees of the acquired firm to amend or transfer their loyalties.
- *Joint ventures* – Two or more organizations can be involved and the structure may be formal or informal. Joint ventures are covered by competition legislation to varying degrees in different countries.
- *Knowledge purchasing* – This may be subdivided into two categories:

- *Contract research* – Generally a higher level of integration as the contractor is bound by terms and conditions to deliver certain results/information (knowledge). The research contract may be similar to the employment contract, for example specifying exclusivity and confidentiality to outlast the contract itself.
- *Licensing* – Here the knowledge buyer licenses a right to the technology or information required. This is seldom an exclusive licence and therefore other organizations (competitors!) may be integrated in the same way. The licensor maintains control over the licensed knowledge.
- *Knowledge scanning* – Probably not covered by a contract. Legal and illegal forms may be encountered. May be less expensive but results in a lower level of knowledge transfer.

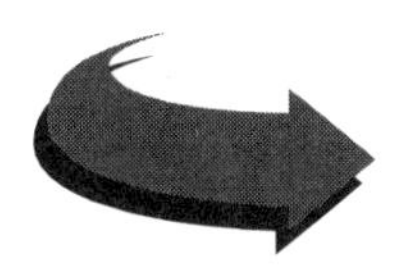

Of particular interest to this book is knowledge buying, but it is useful to see knowledge buying in the context of other sourcing strategies. The first question for the knowledge-hungry organization is, do I need to buy this information or can I acquire it in another way?

IS EXTERNAL KNOWLEDGE ACQUISITION INCREASING?

The Granstrand study mentioned above included compelling evidence of the increase in external technology acquisition in Sweden and other countries. A Contract Research Survey carried out in the UK in 1998 gave a clear message from R&D practitioners that *contract research* is increasing. Anecdotally this appears to be the pattern in other countries. Many organizations are finding it necessary to increase the width of their knowledge base because of technology diversification within their markets, increased regulatory requirements or the demands of customers, each of which can result in need for greater knowledge and/or greater technological competence.

External knowledge acquisition is especially valuable in the context of complex new technologies requiring expensive R&D, as well as significant time, if developed internally. Organizations may be unable to find sufficient in-house competence even where they can afford the time and expense of internal knowledge development. There are some real synergies between in-house and external research: new information from external sources about technology feasibility, without necessarily revealing 'how it was done', can act as a spur to directing in-house R&D.

STRATEGIC CHOICES FOR ORGANIZATIONS

Most organizations today face pressures in keeping abreast of new technological and other developments. Globalization has compelled commercial companies to become more effective in developing new products and processes as increasingly fierce competition threatens all markets. Technology is becoming ever more complex, product life cycles are shortening and substitute technologies follow in rapid succession. This is true of every industry and every activity that has a technology/knowledge basis. Companies are finding it increasingly difficult to carry the internal resources to sustain R&D capability to cover the whole gamut of technology. Many are therefore concentrating their knowledge capital and research into core competencies, and forming strategic partnerships or contracting out research of non-core activity.

It has been noted that *purchasing knowledge services* is a form of integration of the buyer and the seller, albeit on a temporary basis. While this concept is useful, there is a danger that it can be taken too far. There need to be clear boundaries in the minds of buying and selling organizations, especially where individuals are working very closely with their opposite numbers. Organizations must avoid the danger of too cosy a relationship being established, to the detriment of forming other useful (competing) relationships.

The decision to purchase knowledge-based services should be based on the overall strategy of the organization. Evidence suggests this is not always decided in a systematic and closely considered way. Two alternatives strategies illustrate the question:

1. A confidential strategic project may involve close scrutiny of the market place, an allocation of in-house resources to undertake key elements and provide overall management of the project and an informed decision to place a specific discrete package of work outside.
2. A tactical project working within ill-defined boundaries, with a budget for internal and external work, and some generic technical objectives.

In the first example the approach to the market will be structured and disciplined; in the second, the budget holder may be open to undue influence by the players in the market. Neither approach is fundamentally right or wrong but will be influenced by the company's strategy towards its knowledge base.

Whichever strategy is adopted – and the actual strategy will in practice be a hybrid tailored to the needs of the organization – the knowledge seller must convince the prospect to become a client. A great deal of problem definition is achieved through the negotiation process which normally covers both technical and business/contractual matters. The client's principal negotiator will normally be the manager who will have day-to-day contact with the knowledge seller and/or his manager. The negotiator must be convinced and supplied with sufficient data to persuade their manager(s) in turn of the value of any proposed contract. These negotiations will be protracted, typically from initial discussions to contract placement a period of six to nine months.

The knowledge seller must demonstrate that it is both a source of knowledge/skills or technology and a diffuser of technical know-how by:

- having the personnel and track record that indicate a strong probability that it can deliver the project
- adapting technology to the client's needs
- providing a focal point for client liaison
- cross-fertilizing knowledge to meet the client's requirements.

The client must be satisfied that the seller can act as an effective knowledge broker with sufficient contacts in the 'real world' to identify and access opportunities for further knowledge acquisition. The seller must be able to demonstrate an ability to convert those opportunities in a way that can be digested by the client.

INTELLECTUAL CAPITAL

In 1997 Cisco Systems, Intel and Microsoft had combined sales of $34 billion and a combined market capitalization of $250 billion. Ford, Chrysler and General Motors in the same year had higher sales but a combined market capitalization of only $103 billion. This requires some explanation, because it underlines the value that is today attached to intangible assets. The simple question we are entitled to ask (and which the investment markets must ask daily) is does the market overvalue intellectual capital and what value *does* it place on brain power?

It should be stated straightaway that at the time of writing (2004) there is no universally agreed definition of intellectual capital, although the accounting authorities in the major trading blocks are making progress towards this goal. The US's Securities and Exchange Commission (SEC) commissioner Steven Wallman included in his definition of intellectual capital not just human brain power but also brand names and trademarks, even assets booked at historic costs that have transformed over time into something of far greater value – such as an old property acquired decades ago that is now prime real estate but potentially valued at zero on the balance sheet.

Swedish insurance firm Skandia, which has done pioneering work in the field of valuing intellectual capital, suggests that investors should look at intellectual capital in the way we look at a tree. Everything we see in an annual report and accounts is like the trunk, branches and leaves of a tree. It's all the above-the-line activity that makes the company what it is. But this is not the entire tree. Half of a real tree lies underground – the root system. Whilst leaf colour and fruit flavours and size indicate something of the current health of a tree, an examination of the root system may reveal a different picture. Perhaps rot, dwindling water supply, environmental pollution, all of which can change a tree that is healthy today into one that is dead in a few years. So with intellectual capital. Careful examination of these assets can provide useful early warning of emerging problems for the organization.

Skandia considers intellectual capital to be:

1. *Human capital* – the combined knowledge, skill, innovation and employees' ability to meet current work commitments. This includes company culture, philosophy and core values. Human capital in a real sense cannot be 'owned' by the company.
2. *Structural capital* – patents, trademarks, organization structure, hardware, software and databases – and all other elements that support the human capital as it delivers value.

Thus:

Human capital + Structural capital = Intellectual capital (or IC)

Intellectual capital may be a relatively new concept, but in practice it has been around for decades as a common sense reality. Put simply, it is normally considered to be the difference between book value and market value. Readers who want to get to grips with the issues of IC are recommended to access the seminal work in this field: *Intellectual Capital* by Lief Edvinsson and Michael Malone.[3] The authors believe that organizations must have better devices for measuring knowledge, and their ability to create it and convert it into profits. The balance sheet has for six hundred years assisted companies to track tangible assets, but simply fails to capture the value of intangible assets such as new product ideas, impressed customers, or a creative workforce. Recent stock market research demonstrates that the relationship between share price and the reported value of firms equity has weakened in recent years – and this means (says Edvinsson) that annual reports and accounts are less and less relevant as indicators of value. To overcome this problem many accountants attempt to measure more and more elements of the organization's activities, such as the value

3 L. Edvinsson and M. S. Malone, *Intellectual Capital*, HarperCollins, 1997.

of past investments in brands, training and R&D. Accountants find it very difficult to put an accurate value on such things. To do this effectively they would have to estimate future profits from a particular asset, a job better left to markets.

To avoid this problem Lief Edvinsson, who has done much pioneering work in the field of IC, takes a different approach. Rather than attempt to value specific 'assets' and then adding them to get a total, he assumes that a firm's market value is a good estimate of its true worth. He then tries to allocate that value to different categories of assets according to their relative importance to the particular business. Edvinsson defines IC as the difference between a company's market value and its 'financial capital' – the net amount that could be realized by selling its physical assets. IC in turn can be subdivided into human capital (the value of training, experience) and structural capital (the ability to make a return from those trained people). Edvinsson then subdivides structural capital into the value of patents, trademarks, loyal customers, product ideas, business processes and so on. Plainly these are only estimates, but by relying on market indicators, managers can assess their relative position and build a model to evaluate their business more rigorously.

It is important to have a general understanding of the growing importance of intellectual capital to organizations – especially businesses.

Organizations should attempt to value new knowledge to be procured from external knowledge vendors. This should inform the financial business case as management decide whether to invest in external knowledge or to attempt to develop the knowledge internally.

As organizations consider strategies for knowledge acquisition, they typically evaluate the different types of knowledge required to meet day-to-day operations. Figure 5 suggests a basic difference between 'new' knowledge and 'time-honoured' knowledge, both of which need to be kept up to date.

Different types of organization maintain different types of knowledge base. For some, the investment base of knowledge must of necessity be large. Others will access their new knowledge via external knowledge brokers and stakeholders. An illustration of the different types of organization and their typical knowledge-base profile is given in Figure 6.

New knowledge must be learned, internalized and discarded	Time-honoured knowledge, to be adapted to today's realities
Quality	Law
Regulation	Accountancy (profit and loss)
Systems	Ethics
Processes	Profit motive
Existing competition	The law of property
Future competition	
Technology	
Customer expectations	
The state of the art	

What type of knowledge do you need to buy? New, or time-honoured?

What is the shelf-life of the knowledge that you need to acquire?

Figure 5 The constantly changing business environment – new and old knowledge

Multi-technology companies	Start-ups
Central and local government	News media
Military	Regulatory authorities
Medical	Small to medium enterprises
Consultancy firms	Immature industries
Universities and academic	

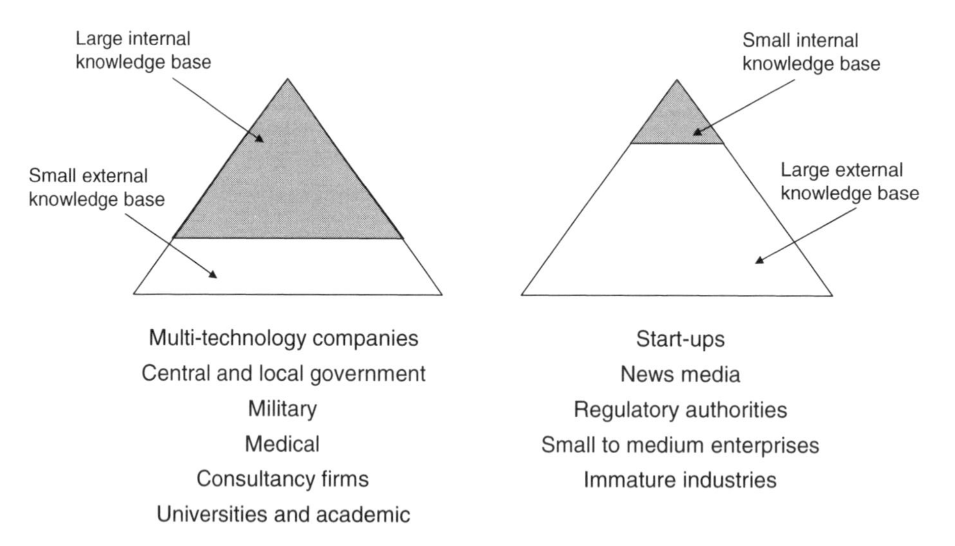

Different types of organization face different challenges as they buy in knowledge from external sources

Figure 6 The size of the knowledge base

As managers plan to acquire external knowledge, a simple yardstick is required. Will the new knowledge be truly new (novel, innovative or otherwise leading-edge) and/or complex? It should be possible to plot a knowledge concept on a simple quadrant to illustrate the severity of the knowledge-acquisition test that lies ahead. This is suggested in Figure 7.

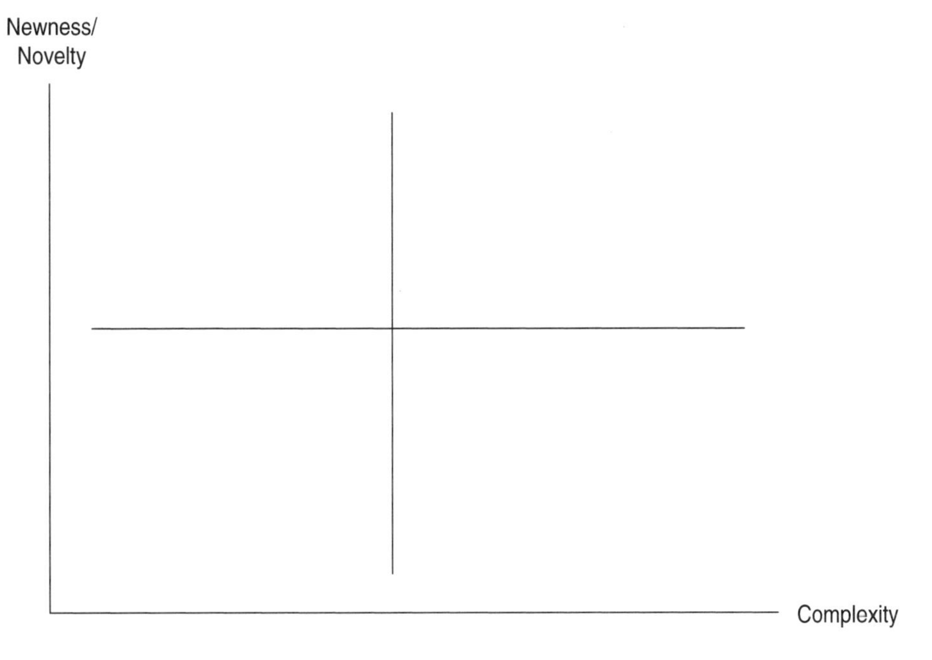

- Buying in knowledge?
  - Where on the quadrant are you aiming?
  - What different strategies are appropriate for each box?

Figure 7 Knowledge types – knowledge targeting

CHAPTER 6

Planning to Buy Knowledge

DUE DILIGENCE ISSUES

In this chapter, attention is given to *best practice* procurement of knowledge. Readers may wish to use this as a benchmark of whatever procedures are currently used within their organization, adopting and adapting as appropriate. The chapter follows the process in a logical sequence from need identification, through requirements definition, invitation to tender (ITT), negotiation, business case development and contract award. Plainly, not all these elements will be present in every case, but it is a useful theoretical starting point to examine the process as it impacts *knowledge purchasing* (see Figure 8).

Previous chapters have suggested some background strategy issues that need to be considered. At this stage it is assumed that the client organization has evaluated alternative options (in-house knowledge development, joint venture R&D, collaborative work and so on), determined its strategy on IPRs, (ownership, background and foreground rights) and is now embarking on the knowledge 'purchasing' process as the best option in the circumstances.

Organizations wanting to undertake a thorough review of their services procurement activity may wish to refer to *The Outsourcing R&D Toolkit*,[1] which contains a very detailed treatment of procurement aspects of contracting for R&D services, a subject that bears many similarities to purchasing other knowledge-based services.

The concept of *due diligence* is of increasing importance today. Whilst it has a specific legal meaning, normally in relation to the officers of a company, the concept has been expanded to include all those things an organization should do in order properly to discharge its obligations to its various stakeholders. In practice this means ensuring as far as is reasonably practical that desired outcomes will be achieved. In the field of acquisition of knowledge, due diligence will equate to properly and adequately defining the work to be done (knowledge to be delivered), making sensible checks of the proposed supplier and then ensuring that any resulting contract is properly managed, in accordance with modern best practice. Due diligence is relevant not only to companies but to all organizations and the individuals that work within them. In this chapter we have broken down the due diligence process into those activities

1 P. Sammons, *The Outsourcing R&D Toolkit*, Gower, 2000.

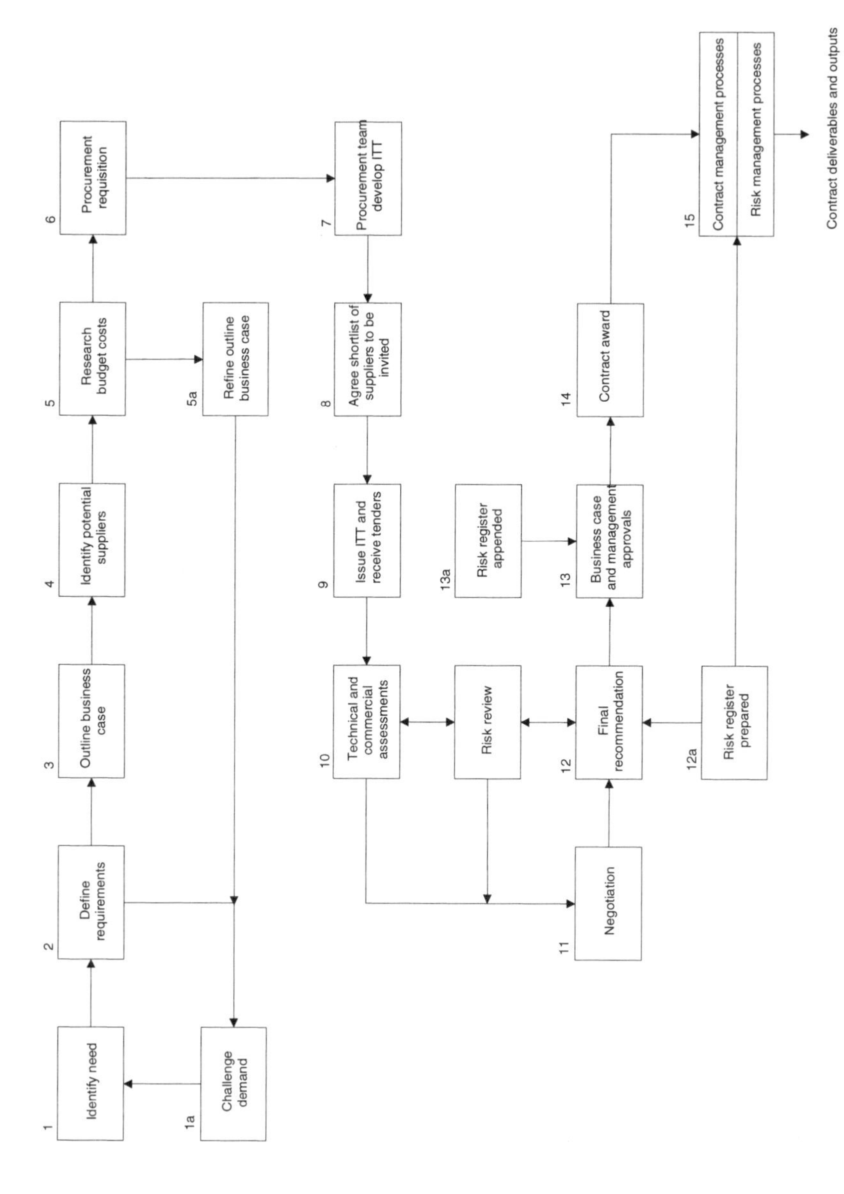

Note: Stages 1–5 are sometimes called the project assessment phase.

Figure 8 Buying knowledge – a generic procurement process

required both before a contract is entered into and then the activities to be done whilst a contract is underway.

VENDOR ASSESSMENT

In any buying and selling situation, each party makes a provisional assessment of the other's status as a business partner. For the buyer the standard questions are, will this supplier be a reliable partner, delivering what I want, when I want it, at a price I can afford? For the seller, will this client be a reliable partner, telling me all I need to know in order to satisfy my order properly? Will the client become a nuisance during our relationship and will they pay me on time?

In the field of commercial purchasing, seller or vendor assessment is an important task for the buying organization. An organization buying knowledge must evaluate the potential supplier before committing to any substantial contract. So what does a client organization need to know about a potential knowledge supplier and how do they go about obtaining the necessary information and data? Indeed, who should undertake this work? To answer these questions in reverse order:

Who should undertake vendor assessment?

This is an activity to be undertaken in a systematic and methodical way. The factors to be evaluated are the technical, technological and professional credentials of the potential supplier, its facilities, management competence/track record and its financial stability. It is strongly recommended that this is a team effort, unless the knowledge buyer has available an individual of exceptional breadth and depth of experience covering both technical and commercial activities. This activity should be overseen by a senior manager to ensure it is given the attention it deserves.

Who presently undertakes this work? What is the optimum mix of skills to carry out vendor assessment for your business?

How does the team go about collecting the necessary information?

This is a matter both of searching for information and expecting the supplier to be candid with information to enable the evaluator[s] to do this job properly. Obviously, if the supplier is hungry for work or believes they may lose business they are more likely to cooperate.

Keep the potential supplier hungry!

Caveat emptor! Although the potential supplier may have a widely known reputation, what is its track record in your particular area?

Assessment is an ongoing process. At least three stages are identifiable:

1. *Pre-contract* – buyer sets out strategic requirements. Potential supplier produces a proposal or agrees to the buyer's specification of work to be done.
2. *During contract* – buyer monitors technical and administrative performance.
3. *Post-contract* – review achievement of objectives (joint review or internal review – or both).

It is necessary to ask probing questions at each stage:

- *Pre-contract* – What do we want from the knowledge to be purchased? What alternatives do we have? How will we use the knowledge?
- *During contract* – Do we wish to change the objectives? Are we achieving the objectives? Should the work continue? How is the supplier's team performing – are personnel changes required?
- *Post-contract* – Did we succeed? Did we get good value for money? Did we utilize the knowledge? What if we had done nothing?

These stages are, as suggested above, analogous to supplier assessment and monitoring in any other purchasing situation. A pre-contract 'visiting group' may be constituted to carry out a detailed assessment of work planned and progress made – it looks at the organizational mission or objectives of the knowledge vendor and at its past achievements. It examines quality and productivity of the potential supplier, at its internal monitoring procedures and may supplement its findings with peer reviews. Typical assessment techniques of specific relevance to knowledge procurement are:

- visiting group/peer review
- level of external funding – if a not-for-profit supplier
- extent of collaborative work

- staff profiles (age/turnover and so on)
- measures of efficiency (overheads per employee, productivity per employee, project completion rates, publication review)
- patent and royalty review
- customer review.

Most of these reviews are self-explanatory and merely require the client to devise the necessary format of question appropriate to the situation.

Vetting potential suppliers as potential information security risk

Preventing valuable information from being lost, leaked or misappropriated is difficult: industrial espionage is a growth area. The R&D activity of a company can give vital clues as to its financial strength and the future products or processes which it may introduce. It can also give an early indication of strategic policy changes and so is a key area of investigation for those involved in systematically assessing their competitors.

Security is a sensible area to review when undertaking a supplier assessment exercise. Ask to see a copy of the potential supplier's security policy, quiz their staff as to their understanding of their policy, see how it is implemented in practice, ask about other work that they are doing at the present and see how much they reveal. Obviously this need not be done in too overt a manner (not on the first visit) and nor need the potential supplier's failure in this vital area necessarily preclude it from working for you. It will, however, place you in a good position to discuss with the potential supplier any security risks you perceive and encourage you to work together to identify and implement any necessary improvements to security.

At the vendor assessment phase you are primarily interested in identifying the potential supplier's culture in terms of information management.

Information sources

There is a range of information sources available to the assessment team, which should decide at the outset which sources will be used:

- the knowledge seller's own client register, if available – take references;
- the knowledge seller's main subcontractors, if any – ask for their main point of contact and explain that the purpose is to evaluate the interaction of those parties who may be working together on your project;

- government, local authorities and professional bodies' information departments;
- the knowledge seller – use of a standard questionnaire is suggested;
- trade press, company search agents, Dun and Bradstreet service and so on.

Technical appraisal

Information required will vary from project to project. The following categories are suggested:

- *Capability*: management, technical resources, quality control procedure, parent company involvement/support, location, industrial relation records, safety policy.
- *Present capacity*: can the potential supplier undertake the work properly bearing in mind constraints of current workload? Consider plant, equipment and staff availability.
- *Related experience*: what has the potential supplier done that is equivalent? Has the potential supplier worked for you before?
- *Availability*: can the potential supplier meet our planned programme? What guarantees are they prepared to give?
- *Finance*: unprofitable or under-financed suppliers may be a source of problems to the client. In extremis there may be a case for obtaining a bank or other guarantee, or having advanced funds held in a trust arrangement to obviate misappropriation of funding.

The potential supplier's financial standing will influence the client's attitude towards financial damages in the event of breach of contract and also to the return of sums advanced.

There is a strong case for disregarding any potential supplier that appears to have financial problems irrespective of its technical merits.

COMPETITIVE TENDER OR NEGOTIATION?

This is a basic choice that confronts every 'buyer' in any commercial purchasing situation. Each has recognized advantages and disadvantages. In the field of knowledge-buying and management consultancy, however, there is often a reluctance to utilize true competitive tendering, partly because it is felt that this is too blunt and grubbily commercial an approach to take in a situation where the knowledge buyer seeks to utilize the very best resource available and where the field of expertise may be very small. It is also, in part, because many management professionals, who are effectively the 'knowledge buyer' on behalf of the client organization, believe that they know the field sufficiently well to identify the best potential supplier without testing the market. They believe that the decision is already made and, in effect, cannot be influenced.

Readers are encouraged to look creatively at whether competitive tendering is a suitable vehicle for gaining competitive advantage, even where this method may have been rejected in the past.

ADVANTAGES OF COMPETITIVE TENDERING

- Competing firms know, unless *post-tender negotiation* is allowed under the tendering rules, that they must tender their optimum bid at the outset in order to progress through to more detailed consideration (under some competitive tender arrangements, a contract may be awarded on the strength of the receipt of tenders with no further evaluation). This arguably means that the best technical and commercial 'fit' emerges early in the overall negotiation process.
- It encourages competitors to be creative in suggesting alternative commercial and technical solutions to the client's requirement (providing the rules of the competitive tender expressly encourage alternative bids).
- In principle a competitive tender against the client's technical specification gives the client some confidence that he or she is comparing 'apples with apples' in evaluating the tenders.
- This can be a very fair way of choosing between competitors in principle, denying any advantage to one over another.
- May make the potential supplier more amenable to trading on the terms and conditions that accompany the ITT document – in other words, it strengthens the buyer's negotiating position.

DISADVANTAGES OF COMPETITIVE TENDERING

- If the client organization selects the wrong group of tenderers, it is likely to get a suboptimal solution to its knowledge requirement.

- Where handled incorrectly, a competitive tender may create or reinforce a climate of expectation in the mind of the client that is suboptimal in terms of its requirements.
- Collusion between tenderers is possible, especially if an open pre-tender meeting has been held.
- The process is time-consuming and expensive, particularly to potential suppliers who must prepare a detailed bid with a less than even chance of winning the work.
- Selection of the same group of bidders on a number of occasions may self-perpetuate the 'victory' of one over the others, leading to lack of commitment on the part of the others.
- It may discourage the best companies from bidding if they feel they have little chance of winning the tender.

The most potent argument against the use of competitive tendering, and in favour of a negotiated approach to contract for buying knowledge-based services, is that competitive tendering militates against building a long-term relationship with contractors. So-called 'partnership sourcing' has obvious advantages in a situation where the knowledge buyer wishes to build the interest and commitment of the potential supplier, encouraging it to identify with the client's needs, limiting the number of suppliers through whom the client's IPRs may leak, and enhancing useful personal interaction between the client's and the supplier's technical teams.

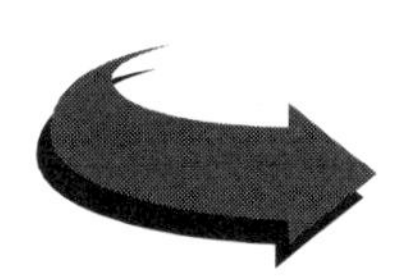

The decision between competitive tendering and negotiated contracts is one to be made at the project assessment phase and is associated with the company's overall policy and strategy on purchasing knowledge services.

Confidentiality of information prior to issuing tender enquiries

One aspect of competitive tendering that is more complex in a knowledge-buying situation than in ordinary commercial settings is the need for confidentiality. Most ITT documents will have confidentiality undertakings incorporated, but this is of little use where, for example, a prospective tenderer declines to make an offer – effectively they will have seen your specification but will be under no obligation to secrecy.

A partial answer to this problem is to enter into a confidentiality agreement with prospective tenderers in advance of their receipt of an ITT document – in other words, if you do not sign up, you are denied the opportunity to bid. Indeed the preceding vendor evaluation process may be made subject to a confidentiality agreement under which the potential supplier agrees not to misuse information about the prospective client to which they are made privy. The problem with all confidentiality agreements, however, is that they are difficult to enforce because proof of information leakage is often impossible. Once information exists, it is likely to leak sooner or later.

A pre-contract confidentiality agreement may be a suitable vehicle to identify and ring-fence the potential supplier's pre-existing intellectual property rights that it wishes to exclude from any rights to be acquired by the client.

SPECIFYING THE WORK

The format of the specification of work will depend on the information to be imparted by the client to the supplier. In the interests of confidentiality there may be a case for preparing two specifications, one in sufficient detail to invite and receive tenders and the other suitable as the basis of the contract. Either way, the specification will cover:

- background to the requirement
- statement of work – scope of work
- timescales or milestones
- project management/administration
- knowledge-transfer targets to be achieved
- documentation to be provided
- technical liaison.

Depending on the quality assurance status of the client organization, which may already have established a required format for technical specifications, a review of the format of specification may prove valuable. If a consistent format can be devised, this will ease the preparation of specifications and possibly establish a useful pattern for suppliers that receive repeat orders.

ESSENTIAL QUALITIES OF GOOD SPECIFICATIONS

- *Completeness*: the purpose of the specification is to define – it must therefore be *specific*. Nothing of importance should be omitted or left to the discretion of the reader.
- *Relevance*: it is important that the specification does not contain extraneous detail.
- *Unambiguousness*: If the client is unclear about what he requires, the specification is not the right place to attempt to clarify his thoughts!
- *Adequacy*: The specification must specify what is required, but not add unnecessary detail. In some circumstances a performance specification, detailing only outputs required from a process, may be better than a detailed technical specification which sets close parameters on all aspects of the work.

A specification is a description in precise terms. It is a part of technical writing which is a branch of literature, not technology. A good specification should normally be intelligible to technical non-specialists, bearing in mind it may well be non-specialists (such as lawyers) who have the final say on the interpretation of the specification! It can be helpful to think of the conditions of contract as being part of the specification, albeit they are normally in a separate document. The technical part of the specification defines what is to be done, the commercial part of the specification (terms and conditions) defines how the project is managed commercially.

TIMESCALES

It is undesirable for a contract to be open-ended as to its period and a timetable of some kind is normally required. This does not, of course, mean that the timetable cannot later be adjusted if the work proves to be more difficult than anticipated or if promising new avenues of work become apparent.

STANDARD OF WORK

The manner in which the work is carried out is ultimately laid down (in an English Law context) by reference to the Supply of Goods and Services Act 1982 which provides that 'In a contract for the supply of a service where the supplier is acting in the course of a business, there is an implied term that the supplier will carry out the service with reasonable care and skill.' The supplier must act with at least the care and skill expected of an ordinary man and, if claiming particular expertise, as knowledge contractors may be presumed to do, then the standard of care and skill rises accordingly.

PROJECT MANAGEMENT/ADMINISTRATION

The supplier will often wish to define its right to manage the project on a day-to-day basis. The client usually will want, for technical and due diligence reasons, to see how its money is being spent and have some influence over the course of the work, especially when knowledge-transfer lies at the heart of the contract. In practical terms it may be appropriate for each party to appoint a project manager, who are the normal points of contact for the two organizations.

CONTRACT PRICING STRATEGY

There is a number of potential contract pricing 'strategies' that may be available to a buyer. The real difference between these contract strategies is the relative distribution of financial risk between the seller and the buyer. To put this in context we will look briefly at the possible range of strategies and then see how these apply in an knowledge purchasing situation. The descriptions used are those found in commerce generally:

FIXED PRICE

Here it is necessary that work can be accurately foreseen/measured and that commencement and completion dates are not subject to change. The advantage is that buyers know their extent of financial liability from the outset, which assists budgeting. Contractual supervision is usually minimal. Disadvantages are that if the contractor hits problems they may be put at risk in terms of losses; this in turn may put the buyer's project and/or programme at risk.

FIRM PRICE

The requirements are as for the fixed price, except that the price, or part of it, is linked to an inflation formula. Useful on longer-term projects as it removes the need for cost forecasting. There is a benefit to the contractor who does not carry the risk of inflation as well as other risks. May require extra administration in terms of cost variation formulae and claims.

SCHEDULE OF RATES (SOMETIMES CALLED TIME AND MATERIALS)

Similar to re-measurement/bill of quantities except that an agreed schedule of labour and materials rates is agreed at the outset. This may be linked to an inflation escalator. Same disadvantages as firm price strategy above, especially extra administration by the client organization.

PART FIXED PRICE, PART SCHEDULE OF RATES

Minimizes the disadvantages as set out above by making a part of the contract price fixed. The buyer tries to maximize the fixed part; the seller may try to minimize it.

COST PLUS PERCENTAGE

Advantages are that minimal specification is required and work can be got underway quickly. The buyer must have high confidence in and trust of the contractor and the contractor is used in helping to plan the work. A 'cost plus' strategy is not often recommended because there is every incentive for the contractor to inflate the cost to increase the percentage fraction (effectively the profit) earned by him or her. Cost plus strategy increases the buyer's burden of supervision.

COST PLUS FIXED FEE

Similar to cost plus percentage, except the fee is fixed, so there is no incentive to inflate costs. There is, however, no incentive to minimize them either, as someone else pays the bill!

NEGOTIATED – OPEN BOOK

This requires negotiation with the supplier, with open accounts and an agreed method of making books available. It requires some financial expertise and a high degree of trust and confidence between the parties. One advantage is that it is helpful to the buyer who in non-competitive situations obtains knowledge of the contractor's pricing methods. Companies in competition with others are amenable to this method as they perceive it as increasing their chances of getting the work. Companies that know they are in a non-competitive situation resist this approach. Another disadvantage is that there is less incentive for the contractor to minimize costs.

THE NEED FOR EFFECTIVE PROJECT MANAGEMENT

We have noted the importance of effective technical and financial control. How is this actually achieved? First and foremost, the Conditions of Contract should support the project management concept. In other words they contain provisions setting out who does what in terms of management and the remedy procedure where disputes or problems arise. The Conditions of Contract or the technical specification should clearly lay down a project management methodology involving:

1. Monthly (or periodic) reports.
2. Named and numbered technical reports to be submitted at key milestones.

3. A 'Project Memo system' to enable tracking of technical and programme communications.
4. On a particularly long or complex project, periodic project reviews held at the supplier's premises to thoroughly review all aspects of the contract – actions arising are clearly recorded and systematically reported in succeeding monthly (or periodic) reports.

This project management methodology is usually set out in the technical specification or appended to it as appropriate. We turn now to review these elements in detail.

MONTHLY REPORTS

These cover:

- Activities undertaken during the month – what has been done, preferably with reference to the technical specification.
- Achievements during the month – milestones achieved, with reference to the technical specification.
- Knowledge-transfer activities – this should be related to the client's specification and show progress towards the client's knowledge transfer requirement.
- Where applicable, subcontract progress – it may be appropriate to append the subcontractor's equivalent monthly/periodic report to the supplier's report.
- A review of Project Memos and any outstanding issues.
- Project Finance report – this may be considered unnecessary on a fixed price contract.

NAMED/NUMBERED MILESTONE TECHNICAL REPORTS

Irrespective of the monthly reports, the project will have a number of milestones set out in the technical specification and these are separately reported. Such milestone reports set out overall project progress and any deviations from the agreed programme. Milestone reports are key contractual events, to which progress payments are likely to be attached. They are named and numbered to facilitate communication at all stages during the project – in other words each report is referred to by its name or number which serves as a focus for both the supplier and the client. Milestone reports cover each distinct/discrete phase of the project as suggested in the following example:

End of Phase 1 – Project definition

End of Phase 2 – Data analysis

End of Phase 3 – Preliminary design

End of Phase 4 – Knowledge-transfer phase 1, 2, 3 and so on

End of Phase 5 – Submission of final contract deliverables.

Milestone reports are *not* used as a vehicle for advising of serious technical or programme delays, or cost increases, although such issues may be referred to. These issues are to be reported in real time and the monthly report is the correct vehicle for this – alerting the client at the earliest stage to known or potential problems.

Milestone reports are formally signed off by the supplier's project manager, normally by signing a signature page at the end of the document. The reports are accepted or rejected by the client organization, normally by the issue of a Project Memo confirming this – see below.

PROJECT MEMO SYSTEM

All communications between the parties on technical and programme issues are sequentially numbered and sent to a single point of contact in each organization, who will:

- maintain a register of incoming and outgoing Project Memos
- distribute Project Memos to any internal addressees
- ensure that Project Memos are properly actioned
- make a monthly report of 'unclosed' Project Memos that are still open, that is, they have not been signed off by the project managers.

For incoming Project Memos the register records the sequential number of the memo and date received, the addressee and the subject. The equivalent register of outgoing Project Memos records the sequential number, the date sent, name of originator, name of addressee and the subject. If this is strictly adhered to there is no possibility of communications going astray or being forgotten. Where Project Memo numbers indicate that some previous memos are missing, these are immediately requested and an effort made to discover why they were received out of sequence.

Project Memos are particularly useful in complex projects where both the client and the supplier have several team members who need to interact with each other on technical issues. Simpler projects should use the same system as it represents good housekeeping and is good training for junior staff on project management techniques. Whether commercial correspondence should be included is a matter for each organization to decide *but with this important caveat*:

> The Project Memo cannot under any circumstances alter the scope or cost of the project or the conditions of contract. For these to be altered the 'contract change' procedure, set out in the Conditions of Contract, must be adhered to.

An example Project Memo is given in Appendix 2.

PERIODIC PROJECT REVIEWS

These face-to-face meetings involve a review of progress to date with particular reference to the technical specification. The following are included:

- Discussion on general progress
- Reports by the supplier's various team members, where the project is a cross-functional one or involves several people doing discrete work
- Discussion on problems and remedies
- Physical review of facilities and work done
- Budgetary review (except on fixed price contracts where this may be unnecessary)
- Agreed actions to remedy problems.

Whilst the project review can be used as a vehicle for deciding that the contract may be amended or varied, the review itself is *not* used as the method to enact such amendments. Both parties should be careful to use the important phrase of 'subject to contract' whilst discussing any tentatively agreed changes. The agreed action list should state that changes agreed to be relevant to the contract will not be valid until a contract amendment document is issued by the client.

CONTRACT CLOSE-OUT

When the project has reached a successful conclusion, it will be necessary to formally close the contract and in so doing, signify that all outstanding liabilities have been satisfactorily discharged (except those, of course, which survive the contract – such as confidentiality). This will be the signal that final payments may be made and that all intellectual property rights, where appropriate, have been transferred. The client organization must be fully satisfied that:

- Contract deliverables have been received
- Both sides have fully discharged their various contractual liabilities
- All disputes are settled
- Documents and materials have been disposed of or returned to the client as set out in the contract
- Where a price variation formula was used to adjust contract costs, that final calculations under these have been made (there is sometimes a delay in waiting for the relevant indices to be published).

Where the contract has come to an end prematurely either through default or for convenience the client must still go through a formal contract close-out procedure.

CHAPTER 7

Knowledge Transfer

When an organization buys goods and routine services, it is usually (reasonably) easy to determine whether, and to what extent, the required contractual deliverables have in fact been delivered. When organizations move into more esoteric areas of services procurement, such as technological or scientific R&D, it becomes rather more difficult to know with certainty how well the services have been carried out. But when the main objective of the work is develop *new* knowledge and *then* to transfer skills and knowledge to the commissioning (buying) organization, the difficulties are magnified still further.

When little Johnny is taken to school for his very first day by his doting parents, he is given over to the care of the school in the fond hope that knowledge will be transferred at a measurable rate and that as the years roll by, steady progress in learning will be achieved. Some thirteen years later when Johnny leaves school for the last time, still doting parents will hope, on the strength of some pieces of paper, that Johnny now knows enough to hold his own in the world and, perhaps, to continue his learning in the rarefied atmosphere of a university. Johnny will by this time have a belief in the value of the knowledge transferred to him by his teachers over the years. He may even believe (perhaps with justification!) that he now knows more than doting mum or dad.

Schooling seems straightforward. Why is it then so challenging to transfer knowledge into organizations? For one thing, no organization can wait for thirteen years for such transfer to take place. There is always a 'shelf-life' aspect to this: once an organization has decided to pursue a certain course, it needs to get on to achieve its objectives with alacrity. Second, schools are established for one purpose only, to transfer knowledge, and to a large extent a school's bank of knowledge is steady-state and generally not subject to rapid change. In business and government organizations, however, when knowledge and skills transfer takes place, very often some or all of the knowledge and skills are new or otherwise leading edge. This amplifies the problems, especially as time and budget are generally restricted. The 'transferor' may itself have only a few years – perhaps even only a few months – exposure to the new knowledge or skill.

The remainder of this chapter establishes some ground rules for ensuring that knowledge transfer is planned, executed and measured in a systematic way that will give the knowledge buyer confidence that his or her organization's investment will prove effective. First two simple definitions for our purposes:

- *Knowledge transfer* – theoretical knowledge, possibly including ownership of intellectual property rights, to enable an organization to comprehend all the relevant facets of a subject or discipline, transferred from a third party into our organization
- *Skills transfer*– practical hands-on skills to enable the knowledge to be used in a practical way to support the organization's day-to-day operations. Skills transfer is a subset of knowledge transfer.

It is worth adding at this point that there is no such thing as perfect knowledge transfer; there will always be barriers to transfer and/or assimilation. Most readers as managers or other specialists will recognize that it can take many years to acquire a truly excellent knowledge of a subject. Most of us will very occasionally have said – if only inwardly – 'I've worked in this field for [n] years, but until today I never properly understood that!' We need therefore to set realistic goals around any project to acquire new knowledge, especially where the project involves buying in knowledge from external sources.

Caveat emptor! Knowledge transfer is not the same as *experience* transfer!

KNOWLEDGE-BASE ASSESSMENT

Every organization must from time to time carry out a formalized assessment of its knowledge base. There is a sense in which organizations (and individuals) do this intuitively and on a daily basis. Managers will normally attempt to keep up to date with business and world news, as well as developments in the various professional fields. On a daily basis this gives them sufficient information to respond to day-to-day threats and opportunities. It is when a strategic change is under consideration that a more formalized *knowledge-base assessment* will become necessary. Most large organizations have tactical, short-term plans as well as longer-term strategic objectives. The latter should drive the former, but it does not always work out so neatly! Strategic plans are kept under review and significant adjustments will trigger various types of review, of which a formal knowledge assessment will be just one. A schematic flow-chart showing a typical knowledge-base assessment process is given in Figure 9.

The knowledge-base assessment is designed to establish, in organizational knowledge terms, where we are today, where we want to be in a given period of time, what are the knowledges and skills that we presently possess, what knowledges and skills will we additionally need in order to achieve our objectives, and finally what barriers or challenges exist to prevent this. To take a practical example:

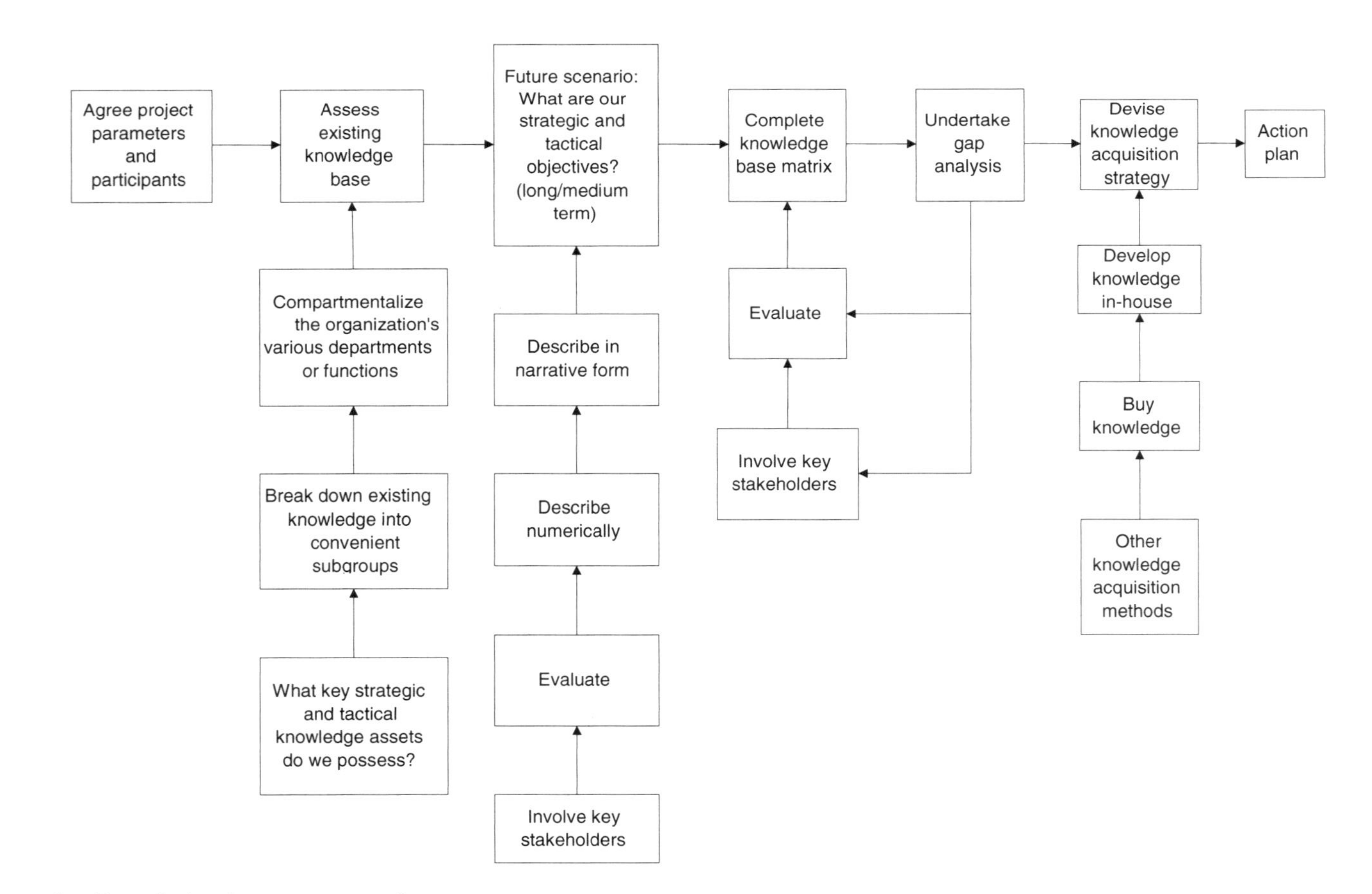

Figure 9 Knowledge-base assessment

> A commercial organization finds that it must change its business processes in order to take advantage of new market opportunities and move away from market segments that have become saturated, with too many players chasing too few customers. To do this there is a recognition that the organization must now address a range of dissimilar markets with a range of products. IT will play a significant enabler role in this, with back-office systems and customer relations management functionality seen as being core to the organization's future operations. So the organization must first establish what skills and knowledge it currently possesses. These will be based around the existing core (or legacy) IT systems and business processes. If this is a commercial organization with manufacturing and/or distribution operations, these too will have a significant existing knowledge base supporting them.

What existing knowledge will no longer be relevant to the future as we envision it? What technology is now obsolescent or approaching obsolescence? What knowledge and skills associated with these technologies will no longer be required in the future? These latter types of knowledge can be allowed to wither on the vine. Our imaginary organization can now envision its end point, in terms of organizational, business process and systems changes. As regards knowledge competencies, it can plot this information as a matrix (see Table 9).

Table 9 Knowledge competencies evaluation matrix

Knowledge needed	*(a) Where are we today?*	*(b) Where do we need to be?*	*Action plan to get from (a) to (b)*
....	....1, 2, 3, 4	.........1, 2, 3, 4	....
....	....1, 2, 3, 4	.........1, 2, 3, 4	....
....	....1, 2, 3, 4	.........1, 2, 3, 4	....
....	....1, 2, 3, 4	.........1, 2, 3, 4	....
....	....1, 2, 3, 4	.........1, 2, 3, 4	....
....	....1, 2, 3, 4	.........1, 2, 3, 4	....

1 = suboptimal knowledge, 2 = basic acceptable level, 3 = thorough knowledge, 4 = excellent, pace-setting knowledge.

The activities necessary to undertake this assessment are twofold: first to compartmentalize our organization into its main constituent parts and list the knowledge and skills that each constituent currently possesses. Second, to visualize and articulate the desired end point in terms of the organizational strategic aims and the skills/knowledge required to achieve those strategic aims. Put at its simplest, where are we today, where do we want to be tomorrow and what are the steps needed to get there?

A useful technique is to describe in narrative form the desired future situation, and then describe the existing situation today. This helps to focus on practicalities as well as give a sense of the scale of the change required. It then becomes easier to visualize the future as well as compartmentalize our present knowledge/skills base.

Self-assessment of the knowledge base is a valuable method, but it does have disadvantages. First, we may underestimate the complexity of the steps needed to move from the present to the future. Second, we may simply not know what we don't know! Self-assessment, however, is a useful exercise before an organization considers inviting an external partner to assist in undertaking a knowledge-base assessment – even if only to validate any future findings and recommendations which that partner may make. It is also true that organizations tend to underestimate their existing skills and knowledge – a deficiency not helped by the fact that a considerable number still have no formal knowledge management programme in place. Accordingly they find it difficult to compartmentalize their organization and recognize their full range of existing knowledge assets. As we assess our existing and target knowledge assets, we should also consider the half-life of the new knowledge we want to acquire. What is its

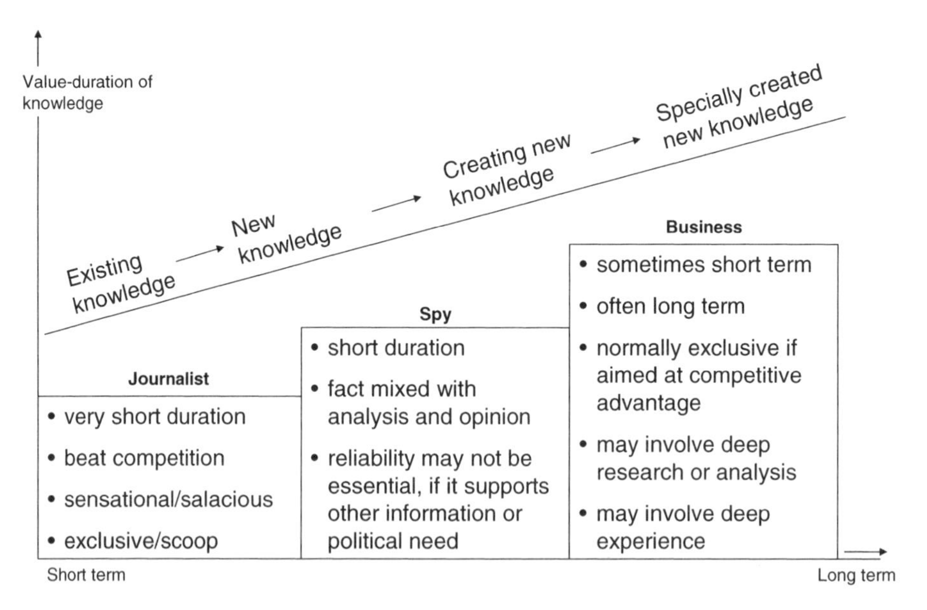

Figure 10 Knowledge types – the half-life of knowledge

present and potential value to our organization? How long will we require the new knowledge to be a valuable, operational asset? A slightly tongue-in-cheek schematic showing the half-life of knowledge to different types of organization is provided in Figure 10. It indicates that knowledge assets have very different life cycles for different types of organization.

A tough question: should a consultant partner be used to help with the knowledge assessment exercise? In considering this we need to reflect on how complex the problem is and what level of vested interest a potential partner may have. Obviously most consultant organizations have an inbuilt interest in developing an ongoing relationship with their clients, in winning repeat business and in helping to deliver more than one phase of a project. So the consultant partner may want to stay as long as possible – and to exaggerate both the client organization's deficiencies and their ability to remedy them! We will look in greater detail at the use of specialist and management consultants in Chapter 8. It needs only to be said here that at this phase the client organization should think creatively about how to encourage any external consultant to achieve its deliverables and then go! A success fee or bonus may act as a powerful incentive to achieve this. Two examples:

1. Consultants may be incentivized to leave through a *bonus for early completion* (define the end point carefully to include measurable knowledge transfer and the conclusion of the relationship – including final payment).
2. A gainshare arrangement: if the consultant can deliver the knowledge assessment project for less than a realistic financial target at a given end-date, then he shares in the 'gain'.

It is also true that most consultants have a model for effecting change and running projects. The client organization must always ask itself, critically, is this model the right one for our circumstances? Will the consultant, in using this model, deploy the skills that we need at this time? And conversely, whatever model the consultant plans to use, does the client need to retain any elements of this on an ongoing basis? What aspects of the model (itself a knowledge base) will we need to have transferred to our organization? The consultant will eventually take his model away. Will we then sink or swim? How much of his model do we need to acquire to manage our 'business process' on an ongoing basis? It is unlikely that the client will ever need to acquire the whole of the consultant's delivery model, but elements of it may be necessary to support certain ongoing operations.[1]

1 There is an obvious intellectual property rights (IPR) issue here. The pre-contract negotiations must clarify the usage rights over various elements of IPR that will be deployed during the consultant assignment.

DEVELOPING A TRANSFER MODEL

Having undertaken a knowledge-base assessment, we can now establish what types of target knowledge can be acquired cheaply and easily, what can be developed internally and what must be sourced externally. For externally sourced knowledge, this is an activity that will take place at this point in the acquisition process. An organization must also consider what will be the best type of transfer model for its own business purposes. The types of model that can be used range from completely passive to extremely active. When a company lawyer buys a new law book there may be no intention of actively reading it, but the knowledge it contains is available to the company with minimum effort, as and when required. This can be thought of as *latent* knowledge. When a scrap metal merchant that specializes in trading ferrous metals takes a strategic decision to move into the non-ferrous metals market, it needs to research, search out and acquire a range of knowledge about non-ferrous metals and their markets – knowledge that will become part and parcel of its future day-to-day operations. This can be thought of as *highly active* knowledge transfer. The knowledge-transfer models available can therefore be plotted as a continuum, as shown in Figure 11.

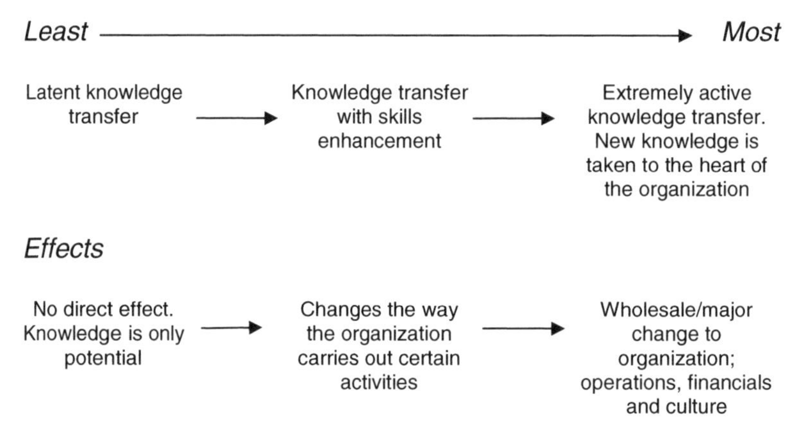

Figure 11 Knowledge-transfer models

A prime objective is for the client to be able as soon as possible to work independently of the external partner. Knowledge transfer is therefore likely to be a critical element to the success of the overall project. The working style that best supports knowledge transfer is a cooperative, team-oriented coaching style – and this is especially true where the partner organization has within its team key personnel with in-depth knowledge of the subject to be transferred. The idea is to leave behind an enhanced knowledge base that meets the new or emerging strategic objectives of the client organization, with knowledge that can be leveraged during and after the knowledge-transfer (KT) project. In order to achieve this the team members must actively partner with each other, as soon as the project allows (i.e. once the contract has been signed[2]).

The following stages are necessary:

1. *Identify knowledge objectives* – the knowledge transfer process begins by identifying the objectives of the KT process. This helps to identify the right level of documentation and the best methods for knowledge capture and dissemination. This activity will provide the client with the ongoing capabilities required once the project has finished. Knowledge objectives are the areas in which the new knowledge is required to be transferred.
2. *Identify participants* – the key individuals who will act as specialists and experts from both client and partner organizations. These people will almost certainly be written into the contract as *key personnel*. For projects where KT is a key element, the client organization may appoint a knowledge transfer coordinator in addition to the overall project manager.
3. *Select enablers* – the team will identify the specific 'enablers' that will facilitate KT. Examples are:
   - classroom training
   - information swap meetings
   - 'shadowing' external consultants with client key personnel (sometimes called the 'buddy-buddy' approach – see below)
   - peer coaching.
4. *Identify metrics and create a template for measurement* – this means identifying the parameters or metrics that will effectively measure and track the success of the KT activity. The template specifies each of the KT methods and helps all the team members to actually manage the KT process by providing a systematic method to track progress. A balanced scorecard approach is one that many organizations use.
2. *Measure the outcome* – formalize the measurement of KT achieved, including not only 'head knowledge' gained by the client's key personnel, but also organizational process changes achieved and, where relevant, behaviour changes engendered. This measurement activity will be enhanced by periodic re-measurement, which is designed to ensure that benefits are realized not only as soon as the project finishes, but are sustained at defined periods thereafter. NB It is suggested that this

2 For sound commercial reasons actual work should not begin until the contract is in place, that is, signed and (where appropriate) accepted.

periodic measurement and re-measurement should extend no longer than 12 to 18 months after completion of the project. Thereafter the metrics may become blurred, and new tactical (or strategic) objectives will then be on the organization's forward agenda.

STAFF SHADOWING

How do these considerations translate into a knowledge transfer model? When the knowledge transfer is only to be *latent*, it is simply required that new knowledge is flagged somewhere in the corporate memory so that those who need to be aware can access the knowledge on the rare occasions it is needed. For the purposes of this book it is assumed that *active* knowledge transfer is the name of the game – that we are buying in new knowledge and skills that we actively intend to use. The knowledge transfer model requires thought and planning, as indicated above. Staff shadowing is one of the key methods to achieve in-depth knowledge and skills transfer. Two major variants are available: both focus on the use of key personnel in the buying and selling organizations.

In their simplest form these models utilize key personnel working together, some from the client organization and some from the external partner. The client requires a person from the partner organization, probably at manager or professional grades, to work with an opposite in the client organization. The objective, simply, is that the client's personnel will receive the benefit of knowledge and skills transfer from their partner opposite. This method is sometimes called the 'buddy-buddy' system. The external partner 'buddies' with his opposite, taking personal interest in transferring useful knowledge to his buddy in the client organization. (It helps enormously if the buddies actually get on together – personality clashes can act as a brake on knowledge transfer!)

It quickly becomes apparent that there can be a number of variations on the buddy-buddy system. A buddy can work one to one with his or her opposite, or one to many, depending on the nature of the work to be done and the knowledge to be transferred. The next question is, at what level of seniority within the respective organizations do these buddies have to be? We might expect the partner buddy to be more senior than the client buddy, in order for the client organization to gain from his or her years of experience. Although this might intuitively seem to provide the best chance of effective KT, it will certainly be the most expensive. There may be opportunities for buddying at roughly equivalent seniority. Or alternatively for use of a less senior partnering buddy matched with a relatively more senior client buddy – it will depend ultimately on the nature of the work and the amount that the client can afford to pay.

To develop this simple model a little more: if the nature of the work is essentially managerial then a good case can be made for buddying with a less senior partner. If the nature of the work is primarily technical, then it is more likely that the partner buddy will be equal to or more senior than the client buddy.

Think carefully about the seniority/experience mix that will best deliver the result you require.

SKILLS TRANSFER

Skills transfer is a subset of knowledge transfer. An employee can know a great deal of theory (head knowledge) but have zero proficiency in the subject at a practical, hands-on level. For example, most middle-ranking business people working in large organizations will have a rough theoretical knowledge of financial accounts and cost accounting: however, that does not mean that their employer will let them loose on the annual report and accounts! Such head knowledge may not even be enough to prepare an accurate balance sheet. A further example: in today's developed world where virtually everyone has some level of regular access to a motor car, either as driver or as passenger, most will have a reasonable grasp of the Highway Code and the mechanics of driving. But few of us would feel comfortable allowing someone with only head knowledge and no practical experience to chauffeur us on a highway! Skills transfer is necessary, therefore, to ensure the payback on knowledge transfer and in most situations of buying knowledge, skills transfer will be a significant part of the client's overall requirement.

The knowledge-base assessment method described above provides an opportunity to identify and evaluate skills status in the same way that we itemize and evaluate knowledge levels. We can then, in the same way, place metrics on the skills we need to acquire in the drive towards our organization's strategic objectives. The two methods, skills assessment and knowledge assessment, are so close that it may be artificial to force a distinction between them. If, however, we do need to make this distinction then the obvious way to do so would be to simply evaluate knowledge and skills separately but use the same mechanism to record them. Hence the evaluation pro forma might be configured as shown in Table 10.

One of the most effective methods to achieve skills transfer where an external partner is being used is through staff shadowing. In a sense this is what happens when a new employee is brought into an organization and there is a handover period, often with the previous job holder taking the new employee through the tasks and responsibilities to be assumed – in other words showing the ropes to the newcomer. In the context of buying knowledge, it is important for the project manager to consider

Table 10 Knowledge and skills competencies evaluation matrix

Knowledge needed	*(a) Where are we today?*	*(b) Where do we need to be?*	*Action plan to get from (a) to (b)*
....	....1, 2, 3, 4	.........1, 2, 3, 4	....
....	....1, 2, 3, 4	.........1, 2, 3, 4	....
....	....1, 2, 3, 4	.........1, 2, 3, 4	....
....	....1, 2, 3, 4	.........1, 2, 3, 4	....
....	....1, 2, 3, 4	.........1, 2, 3, 4	....
....	....1, 2, 3, 4	.........1, 2, 3, 4	....
Skills needed	*(a) Where are we today?*	*(b) Where do we need to be?*	*Action plan to get from (a) to (b)*
....	....1, 2, 3, 4	.........1, 2, 3, 4	....
....	....1, 2, 3, 4	.........1, 2, 3, 4	....
....	....1, 2, 3, 4	.........1, 2, 3, 4	....
....	....1, 2, 3, 4	.........1, 2, 3, 4	....
....	....1, 2, 3, 4	.........1, 2, 3, 4	....
....	....1, 2, 3, 4	.........1, 2, 3, 4	....

1 = suboptimal knowledge, 2 = basic acceptable level, 3 = thorough knowledge, 4 = excellent, pace-setting knowledge.

both the knowledge and skills elements and to make any distinctions necessary. Skills transfer will normally be an important dimension of any knowledge-transfer project, and an important success criteria through which the project's effectiveness will be measured. It is emphasized, once again, that planning this process will make all the difference between success and failure. The following steps should be considered:

- *Team training* – to provide awareness of the skills-transfer dimension of the project. This will include both internal and external team members.
- *Capping partner involvement* – where an external partner/consultant is used to assist in developing or delivering the project, place limits around this involvement. Fixing the price, rather than using a time and materials or cost-plus strategy will encourage the partners to move on at the right point.
- *Establish a skills-transfer subproject* – this will include project team members and operational staff who will be utilizing the new knowledge and skills. If there are particular skills held by any external implementation partner(s) that need to be transferred to the client organization then these should be identified and a transfer process designed. The skills-transfer project is responsible for testing the effectiveness of transfer at various stages of the project.

EMBEDDING KNOWLEDGE

This section follows on naturally from skills transfer. There is little point in paying out perhaps considerable sums to buy in new knowledge and acquire new or enhanced skills if these refreshed knowledge assets do not become embedded in the organization. What are the symptoms that this might not be happening? There may be a 'reversion to type' where employees, having learned new information, methods and skills, revert to the old ways of doing things. If the task can be done in the old way, it might be simpler for them to do so – even if the new way is intrinsically better (or even easier!). There may simply be a clash of cultures, where the previous knowledge base, the old way of doing things, is deemed to be better. This latter phenomenon is especially likely following significant organizational change such as a merger with a former competitor, a takeover, a major change in status such as privatization of a former state asset, or a change in status from a private to a public company. All these situations suggest a cultural history and (probably) transfer of staff who may, quite legitimately, have loyalties to the old ways.

Sensitivity needs to be applied by management together with specialist knowledge transfer techniques to ensure that staff fully understand the new organizational objectives, reasons for change and so on. Loyalty to a former organization is generally a good thing, but employees may need special help to understand the new environment in which they are expected to work. There are, however, plannable methods to achieve the embedding of new knowledge – whether in an individual and his or her way of working, or in an organization – and the cultural change it must achieve to ensure payback on its investment in new knowledge.

To consider this in the organizational context we can take a brief look at the training of military pilots – training that generally takes students through at least three distinct phases and which can be thought of as knowledge transfer par excellence. When a young man or woman decides to join their nation's air force with a view to pilot training, they will very probably have little or no knowledge of flying. Or possibly, if they have had the benefit of a university education, perhaps under an air force scholarship, they may indeed have acquired a private pilot's licence via university arrangements. Either way they will be trained in flying skills virtually 'from scratch' – from a zero knowledge base to a knowledge and skill base of considerable proficiency. They will, after all, early in their professional careers, be given charge of assets and responsibilities that in other contexts would require them to be at least halfway up the managerial chain of command.

From basic *ab initio* training pilots move on to operational conversion, where the foundational skills they have learned will be refocused into an operational context. Put simply, they transition from people who can 'drive a plane around the sky' into professionals who can carry out missions in an operational context. Finally they are assigned to an operational unit (a squadron) where they acquire further operational understanding, learn about squadron life, history, culture and so on, as well as building up hundreds of hours of routine training that are designed to enable them to respond and react to a host of situations in a reflex-like manner. The three phases of training – of knowledge transfer – can be depicted in a simple schematic as shown in Figure 12.

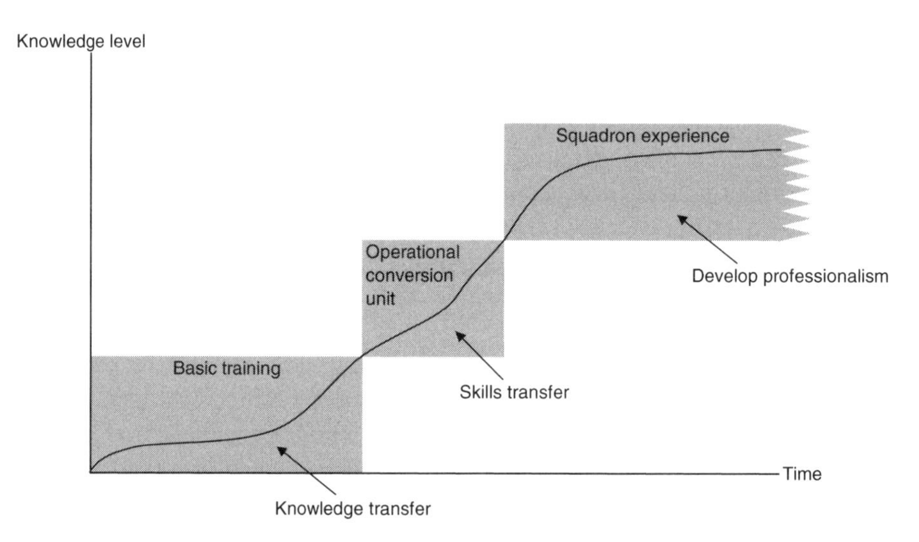

Figure 12 Military pilots – knowledge transfer and embedding

Pilot training of the type described above is necessarily complex, time-consuming and expensive. But it is analogous at a macro level to the stages that are undertaken when an organization at micro level buys in knowledge. These stages are first, *establishing the basics*, second, *contextualization* and finally *skill enhancement*. All organizations need to consider how knowledge once acquired will be retained, used and enhanced. Embedding new knowledge is an aspect of *change management*. As much has been written in recent years about organizational change management there is little we need to add here, except to suggest that as our organization plans the acquisition of new knowledge it must consider how the new knowledge might alter the way the organization conducts its day-to-day operations. We noted earlier in this chapter that the knowledge-base assessment should arise out of an organization's strategic plans – those plans will have set out the organizational objectives at a high level. Strategic change, when it occurs, is likely to impact what the organization does and the way that individual employees do their jobs on a day-to-day basis.

To look at these two areas in turn, there are some basic questions that will be faced in most change management situations.

IMPACTING WHAT THE ORGANIZATION DOES

Organization design – the future strategy

- Will responsibility for major organization functions shift from one department to another?
- Will departments be created, merged or removed altogether?
- What geographical locations will be impacted by change?
- What is the present organization structure across functions and locations?
- Do we have a clearly defined set of task definitions and job descriptions for the areas likely to be affected?

Organization processes – the future strategy

- Will processes and activities flow across separate organization units/functions?
- Will processes be controlled by different functions?
- Will processes be measured using different or new metrics?
- Looking through the various processes at different levels within the organization, which will change and which will remain the same?
- What is the present status of process documentation and documentation control?
- What in-house resources do we have to support process design and/or re-engineering?
- Is there presently a backlog of work or other operational strains?

IMPACTING HOW INDIVIDUAL EMPLOYEES DO THEIR JOBS

Training

- What is the present training capability of our organization?
- Are training materials developed in-house or via external partners?
- What numbers of employees will be affected by change in each Function?
- Has the organization previously undergone change on a significant scale? What was the success or otherwise of that experience?
- Is there any 'change fatigue' currently within the organization?

Performance management

- What is the present performance management system and how does it operate?
- Are performance targets aligned and cascaded clearly? Do employees understand the targets?
- Does the present system support effective target-setting for individuals (or groups)?
- Does the present system support effective tracking of individual performance in line with organizational objectives?
- Are there any other initiatives presently being implemented that could inhibit or impact our organization's ability to adapt to change?

Culture

- How well does the present culture match the vision of the future organization?
- How much experience does our organization have of dealing with change?
- Do we have a record of growth and continuous improvement (assuming these are relevant)
- Do we currently have a network of change agents within our organization?
- What is the past history of change within our organization? Does the presently planned change measure up to past experience of change?
- Are there trade union issues to address as we plan significant organizational change? In the past have the unions supported or opposed change?

If we work systematically through these questions, devising other questions more specific to our particular needs, then our organization will be better placed to devise strategies for embedding new knowledge that has been brought in to the organization. We can then determine not only what are the training and culture-change challenges, but also the sort of communications strategy that will support our objectives. In this way we can accurately assess our readiness to meet the change. Examples of the questions we may need to review are given below:

Communications strategy

- To what extent have we already communicated organizational change strategies within our organization?
- Do we have a clear route-map via which to plan the change process?

- What level of awareness is there at present within our organization?
- What is the extent of involvement, so far, of the various internal stakeholder groups in planning the change?

Change readiness

- How widely is the need for change understood and accepted?
- Can we anticipate the likely receptiveness of various internal stakeholder groups to the proposed changes?
- At what levels within the organization has the future strategy been discussed and agreed?
- Should there be an exodus of key staff as a direct result of the proposed changes, what is the state of the local/regional/national job markets to replenish our team if necessary?

Attention to strategic plans, knowledge-base assessment and planning for the process of change will all enhance the likelihood of new knowledge becoming properly embedded in the organization and an appropriate 'return' on the financial investment and effort put in.

BENEFITS REALIZATION

Even with the best-laid plans and the most attentive and cooperative employees, our organization may still fail to achieve the full benefits anticipated to be delivered by the new knowledge in which we have invested. It is a prime responsibility of the management team to ensure that promised benefits are in fact realized. This is not always an exact science, and there are many different ways of measuring and tracking benefits actually delivered. What follows below is necessarily restricted to benefits realization in the context of buying knowledge.

The first thing to say is that it is essential that as the *business case* is prepared, there is some clear target return on investment (ROI) associated with the strategic plan and with the proposal to source knowledge externally. Good management teams aim to deliver significant changes and improvements to their organizations (breakthrough changes) and to deliver more than just superficial improvements. However, measurement may be hampered by lack of sophisticated measurement tools – indeed a crude yardstick may be the only one available. The author has generally taken the view that a ROI of 3:1 over any agreed payback period is the absolute minimum an organization should target when buying in knowledge. If we cannot achieve this relatively modest benefit, then we should look to other ways to engineer the organizational change or to acquire and develop new knowledge using in-house resources and expertise. There will of course be an *opportunity cost* to diverting our staff from their existing work, and this cost must be factored in to any calculation.

There are four main dimensions to the benefit calculations that support any decision to buy-in new knowledge:

- Return on investment in intangibles
- Procurement process benefits
- Business case benefits – examples:
  - change to business/operational model
  - change to sales lines
  - new revenue streams
  - lower costs – reduced spending
  - revised business processes
- Behaviour change after embedding of knowledge.

It will be readily understood that these aspects overlap and in practice it may be difficult to separate them completely. ROI on intangibles is plainly linked to other business case benefits – which may include lower process costs and/or increased revenue streams. Only *procurement process* benefits are genuinely stand-alone, and in some ways they are the most difficult to measure and probably deliver only the most modest of the financial benefits. Procurement process benefits means two things: first minimizing the cost of the procurement exercise that supports the knowledge buying project and second any financial benefits that arise directly from the procurement activity – simple things such as identifying excess costs and charges in the supplier's quotations, or bringing new suppliers into the project that stimulate competition resulting in lower through-life costs.

A systematic approach to planning the realization of benefits will repay the effort involved very handsomely. The activities, once listed out, are relatively intuitive:

- knowledge-base assessment
- plan benefits in operational and financial terms
- create a benefits list – append this to the business case
- appoint a *benefits realization manager* (this could be the project manager, but it may be better to have someone relatively independent of the project, to bring some detachment to the measurement and tracking of benefits)
- monitor and manage benefits initiatives.

In all too many organizations, the measurement of savings, however defined, has been accompanied by genuine practical difficulties and an element of disbelief by people outside the procurement function. Even worse, financial benefits achieved may be lost on other projects – often referred to as leakage. Any delay to project completion dates can also seriously impact the calculations in the business case.

The business case does not deliver the project. Are your project management and project delivery skills up to the job?

Most business cases will include a financial benefits section, to support the overall investment case. Analysis of financial benefits will focus on implementation costs and post-project benefits, of which the following are the most common.

Savings that have an impact on the profit and loss account

- rate reduction – reduction of supplier's standard charge rates
- reduced leakage – reduced cost arising from employing suppliers on an approved supplier list rather than continually swapping suppliers
- cost reduction – overall reduction in costs
- rebate – linking costs already incurred and business yet to be placed
- demand management – reduction in the extent of work required.

Costs that have an impact on the profit and loss account

- project costs – what it actually costs to carry out a project
- implementation costs – further costs, whether budgeted or not, incurred to complete a project
- ongoing costs – additional costs, not in the budget but incurred as a result of the project.

Savings that provide the organization with other measurable benefits

- cost avoidance – reduction in costs that were incurred prior to the project
- process savings – increased efficiency, reduced expenditure on consumables and so on
- added value services – new services, products obtainable from the project at no additional cost.

Benefits realization calculations should be projected forward for a reasonable and realistic period. It is the author's opinion that projection beyond two years becomes increasingly unrealistic for most companies. An organization will obviously recognize that a significant process change, or other change to the way it conducts its day-to-day operations will benefit the organization perhaps for many years, but in terms of financial planning, within 24 months the benefits should probably be taken for granted (or amortized in some way) and new improvement opportunities should by then be being explored.

What cannot be measured, cannot be managed!

Senior managers should continue to take an interest in, and be accountable for, benefits realization once the practical/delivery phases of a project have been completed.

The overall benefits realization process is outlined in the diagram at Figure 13. This outlines the typical phases through which a project will pass and the likely activities to be undertaken in each phase. Figure 13 will repay close study – if each activity is well planned then the prospects for genuine benefits delivery and sustainability are greatly enhanced.

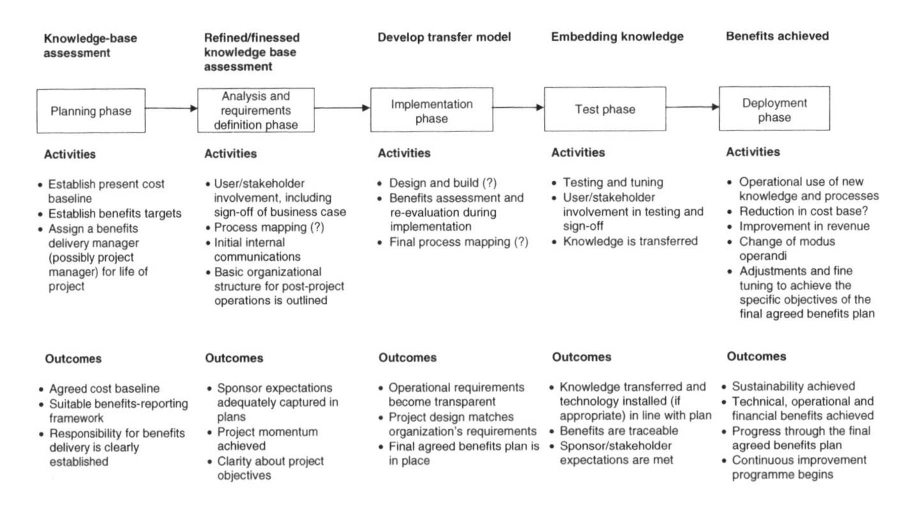

Figure 13 Benefits realization – generic activities and outcomes

CHAPTER 8

Working with Consultants

PROFESSIONAL SERVICES – PROFESSIONAL SERVANTS!

So far in this book we have reviewed the importance of knowledge as the basis of the modern economy, looked at the subject of intellectual property rights, at the need for organizations from time to time to buy knowledge as one of a number of knowledge acquisition strategies, and at some of the practicalities of transferring knowledge from external sources into our own organization. The remainder of the book looks at the specifics of buying knowledge from three distinct knowledge sources: consultant organizations, contract research organizations and finally, universities. In point of fact each of these can be thought of as 'consultants' in the broadest sense and it is true that many of the strategies we are about to review are applicable to each of these three knowledge sources.

Consultants are paid to give advice and sometimes, to assist in implementing that advice. They are not 'an extra pair of hands' – that is contract labour, a subject in its own right. What distinguishes consultants from other forms of short-term assistance is precisely that they advise or provide a package of services of which advice is the foundation and their unique contribution. The centrality of the provision of advice will help to shape both the specification of the services (the terms of reference or brief) and the conditions of contract on which those services are to be provided. In the business realm consultancy is usually provided to management and often directly to senior management.

A useful definition of management consultancy is provided by Douglas Gray:

> Management consultants provide the specialist advice needed when new problems or new opportunities arise that demand skills and experience not possessed to the extent required by the organisation … a consultant is someone who has expertise in a specific area and offers unbiased help, opinions and advice for a fee.[1]

Skills and experience are both forms of knowledge. We might add, then, that a consultant is someone who can bring to bear on a problem the necessary theoretical and practical knowledge that will enable their client to move rapidly to resolve important problems that affect the client organization.

1 D.A. Gray, *Start and Run a Consulting Business*, Kogan Page, 1990.

A prospective client often prefers to negotiate an assignment with the individual who will actually carry it out. Social contact between the professional and the client is important as mutual understanding and trust are stimulated through such contact. Herein lies a significant problem for procurement professionals who may subsequently be tasked to agree a contract and advise on rates to be paid in what has become, in effect, a monopoly supply situation. The decision to appoint a particular consultant and the personal commitment to an individual or particular firm of consultants is often taken before the professional 'buyer' has become involved. This denies the opportunity to inject formal competition into the selection process and weakens the buyer's ability to optimize negotiations in terms of price and contract conditions, whilst also denying the opportunity to systematically compare different consultancy practices and, crucially, what they have to offer to a particular assignment.

Consultants target their marketing effort to the *effective decision-maker* (EDM) in a client organization – often a budget holder – though it must be said that increasingly consultants do recognize procurement professionals as a key client stakeholder in the acquisition process that must be managed along with the EDM during pre-contract discussions. As the scope of a potential assignment may be unclear in the mind of the EDM, he or she may welcome pre-contract discussions with potential consultants to help focus their requirements. This has considerable advantages, but the buying organization must be clear about the downside:

- contractual commitments may unwittingly be given by the buying organization
- undertakings may be given but not effectively recorded and therefore not reflected in the subsequent contract (and may later have to be incorporated via expensive contract variations)
- the EDM may draw so close to the individual consultant through social contact that they become overcommitted to/dazzled by that individual (the 'halo effect')
- the EDM may (depending on the type of organization and its approach to financial delegations) find that this is one of the few purchasing decisions they can make that bypasses formalized purchasing procedures and so be tempted to display managerial virility by making promises that, in the cold light of day, may be found to be suboptimal.

Professional buyers sometimes find that colleagues seek to justify their choice of consultant on the basis that they are sufficiently familiar with the particular professional field to know which consultant is right for the job. This may be correct and appropriate but it runs the risk of making suboptimal commercial decisions based on imperfect knowledge. In these situations the professional buyer may have little option but to rubber stamp such decisions and be denied the opportunity to add real value.

The need for consultant advice often arises in situations of perceived commercial urgency and therefore need to be satisfied in some haste. There may be an attendant feeling that to invite tenders for services and postpone informal discussions will cause delay to the assignment. This is always a false premise and it should be resisted. Any consultant worth his or her salt can enter pre-contractual discussions and/or tender[2] at short notice. It might be added that in these circumstances some senior managers may fear that to invite tenders, to describe the organization's problems to the consultant market at large, could invite personal or commercial ridicule. These managers may therefore wish to hush up the whole issue by inviting only one consultant to bid.

Consultants (especially management consultants) are well aware of these factors and are justified in exploiting them. The knowledge-buying organization must ensure that its managers know of the dangers inherent in pre-contract discussions and of attendant moral obligations. If pre-contractual discussions – or to be more precise, pre-ITT (invitation to tender) discussions – are to take place, the buying organization should give several potential bidders the chance to enter these discussions and use them to help clarify the requirements. It may be appropriate to tell the consultancy firms that at this stage your organization is merely exploring the dimensions of the business problem prior to making a decision on the optimum procurement route. By all means tell them you are in informal discussions with several firms. In this pre-competitive phase you may find that the consultants are prepared, if not to offer free consultancy advice, at least to put some real effort into alerting their potential customer to all the possible opportunities and pitfalls to be avoided. In this phase, if handled correctly, the client organization can build up its understanding of the problem and potential solutions before inviting formal tenders. But the pre-tender phase must be handled sensitively, openly and in a way that will not disadvantage some bidders against others.[3]

2 The term 'tender' here is meant in the English law sense of making an offer to supply that remains open for acceptance until either (a) it is accepted by the *offeree* or (b) it is revoked by the *offeror*. Commercial organizations operating non-English jurisdictions sometimes respond to a client's *request for proposals*, which in the UK would be understood to be a non-binding basis for discussions, by making offers that are intended to be capable of being accepted and thus creating a contract. When dealing with firms in more than one country, it is worth clarifying precisely the basis upon which invitations to bid are made and the type of response expected.

3 The author recalls a senior director of a relatively small limited company telling him about the company's investigations into raising finance to float the company on the stockmarket. The director said that at the time he had little idea – and no experience – of raising such finance. Together with the Chief Executive he visited four merchant bankers, each of which was fulsome with advice. In the first interview they mainly listened. In the second, they recognized common themes emerging. In the third and fourth they were able to ask increasingly directed questions and explore the depth of understanding of the merchant bankers. At the end of this process they felt considerable confidence about launching the project – all on the basis of essentially free consultancy advice.

Managers sometimes stand a little in awe of consultants, who present themselves as experts, able to help you solve your problems. Managers may feel quite inexperienced by comparison and be loath to reject a consultant's advice – especially where they have paid for that advice! Watch the consultant carefully in the pre-contract phase: will they try to take over the project? Will they listen to your requirements or are they working to their own agenda? The whole idea of professional *services* suggests a master–servant relationship. The client may need to remind themselves that (generally speaking) the customer is 'always right' and, whilst we do not want a yes man to be our consultant partner, we do want a consultant who will work within our cultural constraints, and who will recognize that ultimately (no matter how good they are as a consultant) it is the client who pays the bills and the client wants a *professional servant* working for them.

CHALLENGING THE DEMAND

It is very easy, all too easy, for organizations to make a case for expenditure with an external consultant. For many blue chip companies the costs of hiring in professional services soared – in some cases to quite alarming levels – in the first few years of the twenty-first century. This was partly because of growing regulatory burdens. For public sector organizations, growing expectations of ever-improving service delivery, and the attendant political pressures this brings, again means that professional service firms, and management consultants especially, are much in demand. So it is that there is an almost inherent pressure from within organizations to supplement their resources with externals. If for no other reason than that it gives the appearance of making progress! But consultants are expensive, however you look at it. For commercial organizations all external supply expenditure ultimately hits the bottom line. The amount of commercial sales activity required to repay investment in external services is normally significant and demands that the question is answered – is this expenditure really necessary? Before committing effort to contract for professional services, a series of questions should be posed and answered to the satisfaction of the budget manager, if not the board of directors. There are six areas that must be considered ideally *before* approaching an outside professional services supplier.

1. What is the purpose?
   - What is the business/organization objective?
   - Who is the business/organization sponsor?
   - What is the question or problem that needs to be addressed?
   - What benefits are to be secured?
   - What financial or operational value is to be achieved from the work?
2. What is the nature of the work and the priority?
   - What types of skill will be required?
   - What is the expected end result or main output, for example, reports, internal software, designs?

- When is the end result needed?
- When are the critical points in the programme?
- What effect will any delay have?
- Has all work in this area been fully considered?

Any difficulties answering these basic issues will necessitate more work on the requirement definition.

3. Are internal resources available?
   - Have you checked if any of our staff possess the skills/experience to undertake the work? As a minimum check with HR, procurement and the functional area concerned.
   - If yes, are these staff available or can the work be delayed until they are?
   - If no, could our staff be trained to do the work within the timescale?
   - If there is an ongoing need, why not employ someone permanently or on a fixed-term contract?
   - What other internal/external resources might be required and are they available?
   - Are other key internal staff ready and available to provide required assistance or information?
   - Is there a suitable internal manager available to lead the proposed work?
   - If this work introduces a new supplier, what are the ongoing costs of maintaining that supplier?
   - If this work introduces a new supplier, what are the risks of non-performance? (and risk mitigation strategies?).
4. What are the costs and benefits?
   - What are the real costs involved?
     - cost of external spend
     - cost of internal spend
     - opportunity cost – other opportunities that will have to be forgone.
   - Can the benefits be properly quantified?
   - Can the benefits be justified against the proposed cost?
   - What are the comparative costs of using the consultancy resource compared to alternative ways of undertaking the proposed work?
   - Can the work be done free of charge via a 'swap shop'?
   - Does the work lend itself to:
     - no win – no fee
     - gainshare?
5. Are funds available?
   - Is there budget provision available to cover the cost?
   - What other expenditure (planned or unplanned) must also be met?

- What level of financial approval is required?

6. How will we 'internalize' the results of the proposed work?
   - How many people will benefit from the work in new skills acquired?
   - How many people will be able to replicate the work if required again?
   - Are we currently on a level playing field (versus any proposed external professional service provider) in terms of skills?
   - Do we need to nominate one (or more) individual to become the expert in the subject area?
   - How do we diffuse the knowledge/skills more widely to gain added value?

DEFINING THE REQUIREMENT

The specification, terms of reference, statement of requirement or brief must be well defined. Keep in mind:

- What do you want and why?
- Completion date
- What resources (people, materials, research) will you make available?
- What actually needs to be done?
- What external factors need to be taken into consideration? for example liaison and coordination
- What sub-tasks need to be completed before the main tasks?
- How do you want the work done?
- How should the results of the work be presented to you?
- How much can you afford to pay?

Above all, be clear why you think a consultant can help you, communicate the requirement and demand value for money. A well-worn allegation about management consultants is that they adopt a 'cookie-cutter' approach to solving clients' problems. They receive a brief from the client, redefine it to match their preconceptions and then offer 'solutions' based on a 'flavour of the month' management theory. If this sounds far-fetched, note the following observation by William A. Cohen:[4]

4 W.A. Cohen, *How to Make it Big as a Consultant*, AMACOM, 1985, p. 75.

> Frequently you must go back and modify your central problem statement. You may think of a solution that is excellent, but not a solution to the central problem as you wrote it. If you want to include this course of action, you must restate your central problem so that it fits with this alternative.

It is essential to clarify the deliverables and the scope of the assignment. The following elements need to be considered:

- Work objectives – general outline of the 'problem' and the type of advice required.
- Scope and approach – areas, activities, services and so on to be included and the scope to be excluded, where appropriate.
- Method to be applied to obtain the objectives of the assignment.
- Format of results required: specific deliverables. How 'success' will be measured.
- Specific phases or milestones to be achieved.
- Tasks to be performed by the client, including office and secretarial facilities to be made available. Tasks to be performed by the consultant.
- Progress meetings – purpose, frequency, participants and content.
- Output – report, recommendation, software, data and so on.
- Time – anticipated consultant days (if on day/rate basis), proposed start date, required completion date.
- Cost limitation – basis of fees, estimate of total fee and expenses, method of billing and payment terms.
- Termination.
- Authorization of additional work, and change control.
- Confirmation of acceptance.

The brief must tell the consultant what to do, by when and how much it will cost – or any cost or budgetary limitation. The brief must not, however stifle the consultant's creativity. A consultant should be told that they are expected to use their professional judgement in achieving the deliverables and be proactive in directing their energies to meet the objectives. There may be an advantage in keeping the brief short and general. Providing that regular progress meetings are observed and new objectives or deliverables confirmed, the primary responsibility is upon the consultant to ensure that the client is satisfied.

Under Section 13 of the UK Supply of Goods and Services Act 1982, if objectives or deliverables are not clearly stated and agreed in a purchase contract, there is an implied term that the 'supplier' will carry out the service with 'reasonable care and skill'. This places the onus on the consultant to do all that is reasonable to provide a professional and effective service to the client. After all, it is the consultant who holds themself forth as a professional, able to advise the client in determining objectives and assist in achieving them. Under Sections 14 and 15 respectively of the same Act, there is an implied term that where time for performance is not fixed in the contract, the 'supplier' will carry out the service in a reasonable time and where the contract is silent on the question of money there is an implied term that the buyer will 'pay a reasonable charge'. In both cases the Act states that 'what is reasonable ... is a question of fact'.

IDENTIFY SERVICE PROVIDERS

In any buying and selling situation, each party will make a provisional assessment of the other's status as a potential business associate: will this supplier be a reliable partner, delivering what I want, when I want it, at a price I can afford? Will this client be a reliable partner, telling me all I need to know in order to satisfy the order properly, will they become a nuisance during our relationship and will they pay me on time – or at all?

In the field of commercial purchasing, seller or vendor evaluation is an important task for the buying organization. The extent to which such an organization may evaluate a contractor or supplier depends on the resources and budget available to carry out the work involved. Major organizations have permanent and highly skilled teams undertaking this activity. Their contractors or suppliers expect regular approaches for extensive and sometimes intrusive information, responding to which can be a time-consuming task. A much smaller client organization, by contrast, may rely on a buyer's individual attitude to sources of supply, without minimal attempt to assess the current capacity or financial stability. A decision must be made, therefore, regarding the risk associated with the possibility of a contractor or supplier failing to meet his obligations regarding time or quality. Would the consequences of default or failure be sufficiently serious to the client organization to justify investing a prudent amount in pre-tender supplier evaluation to ascertain the present technical and financial position of potential tenderers? What then does a client organization need to know about a potential professional services supplier and how does it go about obtaining the necessary information and data? Indeed, who should undertake this work?

WHAT DOES THE CLIENT NEED TO KNOW?

In very broad terms this problem can be deconstructed to a simple list of information. The client needs to know:

- How long has this company been in existence?
- What is its financial situation – has it the resources to do the work?
- Is it profitable? What are its trends in profitability?
- What is its track record in the technical area of particular interest?
- Who are its main clients? Take written references and follow these up with visits to other clients if at all possible.
- Who are its main suppliers – is it dependent to any extent on third parties? (If so evaluations of other 'prime contractors' may be deemed to be prudent.)
- Who are its key officers?
- Is employee turnover an issue?
- Are there any outstanding legal actions in which it is involved?
- What are its quality credentials?
- What is its overall ethos and culture? Does this match our culture?
- Who are its key competitors? what is the status of its industry generally?

Deploy sufficient resources to undertake this work – especially to obtain other client references.

How does the team go about collecting the necessary information? This is a matter of both searching for information and expecting the consultant organization to be candid with information to enable your evaluators to do their job properly. Obviously, if the consultant is hungry for work or believes they may lose business they are far more likely to cooperate.

Keep the supplier hungry! Keep the seller selling!

WHO SHOULD UNDERTAKE SUPPLIER EVALUATION WORK?

This is an activity to be undertaken in a systematic and methodical way. The factors to be evaluated are the technical, financial and skills credentials of the consultant organization, its facilities, its management competence/track record and its financial stability. It is strongly recommended that this is a team effort, unless the client organization has an individual of exceptional breadth and depth of experience covering both technical and commercial activities, and who is available to do this work. This activity should be overseen by a senior manager to ensure it is given the attention it deserves.

Who presently undertakes this work? What is the optimum mix of skills to carry out vendor evaluation for your business?

Caveat emptor! Although the consultant may have a widely known reputation, what is their track record in your particular area?

ESTABLISH BUDGET COSTS

A surprising number of organizations, including commercial businesses, initiate discussions with external consultants – sometimes authorizing initial work – without any real idea of what the work should cost to achieve. This places them at a distinct disadvantage with potential suppliers in negotiations, not so much because the professional service firm will cheat in pricing the work, but rather because there is greater likelihood of scope-creep, as both parties explore where the correct parameters should be placed around the brief. Research should therefore be undertaken by the client organization to establish budgetary costs – in other words, costs that are moderately dependable and can be worked into a financial business case. Some suggestions for very basic and rudimentary research are provided below:

- Estimate the likely number of labour hours to undertake the work in-house and multiply by internal charge-out rates, or other internal labour rates, as appropriate.
- Contact colleagues to see if equivalent work has been done before. If so, establish the costs.
- Consult internal HR, procurement and operational teams. Look at similar work that has been purchased in the past, or similar tasks that may have been undertaken by in-house teams.
- Contact potential suppliers: when approaching external suppliers, avoid making commitments and, if possible, revealing the budget allocated.

Maintain the view that competition is the strongly favoured solution. Suppliers are adept at conditioning potential customers to accept given levels of expenditure as inevitable. Keep the seller selling!

- Consult your internal finance department. They can supply overhead and accommodation costs, important where work is proposed to be done on your premises. They can also give insights into pricing the work.
- If all else fails, take a roughly calculated guess (a 'guestimate') as to the sort of costs you will encounter externally in undertaking a knowledge buying project.

Should you use requests for information (RFIs) or requests for proposals (RFPs)? This requires careful thought. It can be expensive for suppliers to respond to these requests, but it may be essential as part of the *vendor evaluation* or *supplier pre-qualification* process. You will probably want to avoid duplicating work, however, for both parties. If you follow an RFP with an ITT you may find that much information will be repeated by both sides. Where RFIs or RFQs must be requested, make it clear that no commitments are thereby entered into and that your organization, as a potential client, does not consider itself bound to invite companies that have received an RFI/RFQ to submit a formal tender at a later stage. If possible, avoid causing suppliers to have to bid twice – once in response to an RFI and then again in response to an ITT. It may be better simply to issue an ITT in the first place.

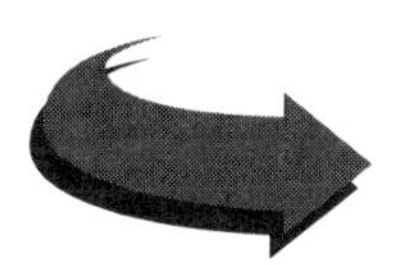

Consider the sourcing strategy, and how you will invite potential suppliers to provide information in the pre-contract stage, as part of the discussions around your overall commercial strategy and contract strategy.

Another difficult question must be considered, one especially important where you are trying to develop budget costs for the potential project: by inviting some companies/organizations to enter into pre-tender discussions, do you inadvertently (or deliberately?) give those companies an unfair advantage over others that will later be invited to tender? It is true that some companies, if they have been asked to provide information in advance of any tender invitation, will have some sort of advantage, if only because they have had longer to consider the problem and potential solutions. This may of course be to the benefit of the client organization, but it runs the counter-productive risk that some potential suppliers may take the view that one of their competitors is almost certain to win the work. In turn they may decline to bid.

Some client organizations take a very strict view of this problem and state that where a professional service supplier has been involved in the *project definition* phase of a project, they are normally to be excluded from the *project execution* phase. This is to avoid any possibility of collusion between the client's staff and the consultant organization. Furthermore, it demonstrably creates a level playing field for other professional services firms to bid for the main *execution* phase of the project. It also means that a client can select a firm with strengths in planning for the project definition phase and other firms with strengths and track record in execution for the delivery phases of the project.

SHORTLISTING OF SUPPLIERS

This will depend very much upon the procurement methods of the client organization and the complexity and value of the consultancy work required. Public sector organizations in Europe – and in other parts of the world – may have a range of special procurement rules that they must observe. If an RFI process is undertaken, this will probably lead to a select list of bidders or select list of tenderers, depending on your organization's terminology. The process of shortlisting may also include the following:

- basic market research
- supplier questionnaires/rfis
- initial shortlist
- formal supplier pre-qualification
- final shortlist.

A refinement of this process, which can be used in more complex projects, is to opt for a two-stage process, whereby a number of suppliers are pre-qualified, perhaps rather more than is ideal, and then a limited ITT is issued, to establish clear financial and technical markers for the work. The limited ITT may use only a reduced outline specification and is likely to elicit promises on consultant rates and perhaps dependable quotes on an upper limit to the cost of the work. By this process you will identify those consultant organizations that are really serious and establish those most likely to be the sort of supplier you want to work with. Some suppliers will be eliminated at the end of this stage and the remainder will be invited to bid against the full client specification, which itself may be developed and enhanced as a result of experience gained via the limited ITT.

INVITATION TO TENDER

An ITT document is a formal way of requesting bids on a common basis. The process is designed to obtain offers from tenderers without collusion, normally using a formalized sealed bid system, by which tenders (legally offers) are delivered on or by a date and time specified in the invitation. Tenders are normally opened at a time and place of the client's choice, sometimes before witnesses.

There is no right number of tenderers to invite. The number must be enough to result in a good spread of bids, sufficient to give a real insight into the state of the market and into the technicalities of the work, but not so many that bidders will conclude they stand only a modest chance of winning and therefore that it is not worth responding. The author normally invites between three and six tenders, depending on the value, the complexity and the market. An in-house procurement group should be able to advise on a case-by-case basis.

The objective is to secure adequate competition whilst still incentivizing the supply market.

Tenderers must be treated on a fair and equitable basis and should therefore be given identical information. The ITT will normally be prepared by an in-house procurement team and usually consists of:

- covering letter
- background information and data
- instructions to tenderers
- conditions of contract
- specification/terms of reference/statement of work
- returnable 'form of tender'.

It may be appropriate to brief tenderers in addition to issuing an ITT. Ideally there should be a single briefing to ensure that all bidders hear the same information and the same answers to questions raised. A briefing provides an opportunity to clarify the requirement and for tenderers to raise questions. A pre-tender briefing must be planned and executed with procurement involvement.

The ITT gives precise information about closing date/time of bid process. A sealed-bid process is desirable for all projects and should be mandatory for larger projects.

Consideration should be given to requesting the technical and commercial parts of the bid to be kept separate to allow separate evaluation of the technical and commercial aspects. Each bid must be kept secure and opened no earlier than the bid return date. In-house procurement teams will normally manage this process. Note that in the UK there are common-law precedents covering the treatment of these sorts of bid process, that require the process to be carried out systematically, fairly and in a way that does not prejudice the reasonable expectations of those invited to bid.

Treat tenderers as you would expect your own organization to be treated!

Tenders must not be invited where your organization has no intention of awarding business.

TENDER EVALUATION

In most cases, certainly in tasks of any value and complexity, it is appropriate to set up a bid analysis/tender evaluation team. It is important that this team has the right level of expertise – commercial, technical, and financial. Legal advice may also be required. The following minimum factors should be taken into account:

Capability assessment

- capability/qualifications of key personnel
- management/supervisory support
- other support systems
- checks on tenderer's references.

Technical assessment

- performance and productivity
- quality
- professional competence
- technical/professional support
- standardization
- after-sales service.

Quality assessment

- quality control systems
- quality certification such as ISO 9000.

Financial and commercial assessment

- inclusion of all relevant costs
- weightings/adjustments needed to make bids comparable
- factors which might affect costs during the contract
- risk analysis
- benefits tracking and realization.

In assessing the relative costs of the various tenders for complex products and services it may be necessary to use a variety of financial analysis techniques, including discounted cash flow analysis, to evaluate the net present value of the proposed work.

The result of these evaluations will provide a method to rank the bids in a systematic way. This may indicate a clear preference/winner, or may indicate areas where further analysis is required.

For higher value and more complex requirements, a formalized risk review will be needed.

Negotiation with tenderers during or after the evaluation may be appropriate. The objective is to seek improvements on the bids, and/or to clarify bids, preferably to the benefit of both parties. Negotiation of course requires careful planning. Advice will be given by the in-house procurement team or commercial department which will normally lead the negotiation team.

AWARD, ENGAGEMENT AND DEBRIEF

It is prudent and important, for reasons of accountability, to record selection decisions and rationale. This record should as a minimum set out:

- summary of competition
- details of competing bids
- advantages of awarding contract to proposed contractor
- implications for existing organization.

A business case is usually the appropriate and logical document for recording these details. The manager or managers with the appropriate expenditure authority should record formal approval in writing before any contract is signed.

There will often be pressure to start the work as soon as possible. It is advisable – strongly advisable – to ensure that the full contract is in place before the work commences. Once the work is underway, the client's negotiating position erodes with each passing day that a contract is not in place. Similarly *letters of intent* are not favoured, but may very occasionally be necessary. Ensure, if so, that these are properly drafted. In the very exceptional circumstances where letters of intent are essential, the full contract approval process must be observed and the signatory should prudently be at one level above that required for the actual contract document.

Signature and (where appropriate) acceptance of a contract gives effect to the engagement of the consultant. The contract should normally be commenced with some form of *contract commencement meeting*. This may finalize any outstanding non-contractual issues, the formal exchange of necessary information, data or documentation. It will also facilitate where necessary, introduction of the consultant to their key opposites within the client organization. Engagement is effectively the end of the contract negotiation phase and leads into the contract management phase that will last until the end of the project.

As soon as practicable after contract award, advise unsuccessful tenderers. Debriefing is best professional practice. Advice on this should be given by the client organization's procurement department, which may lead the debriefing exercise. The reasons why a tender has been rejected are one, or a combination of two key factors:

- it fails to meet a mandatory requirement of the ITT;
- it passes the minimum evaluation criteria but is not ranked as offering the best overall value for money (that is, not ranked as number one).

Although unsuccessful tenderers must not be given the prices of competitors, a general indication of price competitiveness is appropriate in any feedback. Some organizations give details of the range of price bids received, from lowest to highest, so the unsuccessful tenderer can make some estimate as to how it performed in the costing element of their tender.

DELIVERY

Delivery of the contractual obligations by the consultant to the client organization is often the responsibility of a cross-functional team of the client organization's staff, under the control of a project manager or contract manager – hereinafter referred to as the delivery manager. The delivery manager role should be defined in the contract.

The role of the delivery manager is to:

- ensure delivery of cost-effective, reliable service
- ensure delivery of the *contract deliverables* set out in the contract
- manage day-to-day aspects of the business relationship
- seek continual improvement.

The consultant organization's reporting obligations should be set out in the client's specification. A very detailed reporting plan may be one of the initial deliverables under the contract. The following outputs are normally measured:

- quality – compliance with quality management processes
- cost
- time – measured against service levels, milestones
- communications – regularity, timeliness
- benefits realization
- benefits tracking
- skills transfer.

A regular problem with consultant organizations is the client's willingness to add scope to the original brief – sometimes called scope-creep. It is obviously in the consultant's interest to be awarded follow-on work. In terms of buying knowledge, the client's primary objective is to secure knowledge transfer and then discontinue (or 'disengage') the work. Using all the techniques suggested in previous chapters, the client's objective can be thought of as a skills-transfer profile, where the 'partner' or consulted organization's involvement declines over time, whilst the client's

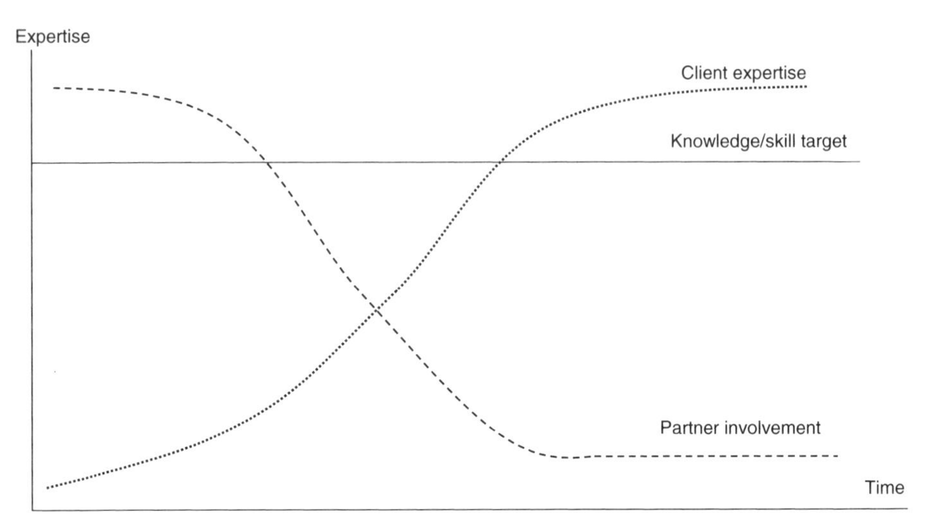

'He must become greater; I must become less' (John, Chapter 3, verse 31).

Does your external partner recognize that your prime objective is to become knowledge sufficient as soon as possible?

Figure 14 Optimized knowledge/skills transfer – from partner involvement to client expertise

expertise/knowledge progressively builds over the same time frame. This is illustrated in a simplistic way in Figure 14. Keep this simple illustration in mind as the project progresses.

DISENGAGEMENT

A formal close-out meeting should be held once the consultant assignment is complete. It is important that this is not conducted solely against an audit checklist. There should be input from both parties with discussion on measures to be used for benefits tracking, where this has not already been decided. The agenda should cover:

WERE CONTRACT DELIVERABLES ACHIEVED AS SET OUT IN THE SPECIFICATION/TERMS OF REFERENCE?

- Were the objectives delivered?
- Were they delivered on time?
- Were costs controlled within budget?
- Were performance standards and expectations met?

WHAT LESSONS CAN BE LEARNED?

- Try to gain feedback from the consultant organization about your organization as client.
- Could we have been clearer in our objectives?
- Could we have helped the work to run more smoothly?

Other issues

- Issues to be determined on a case-by-case basis.

Next steps

- Benefits tracking – refer to Figure 13 in Chapter 7, p. 98.
- Post-project evaluation.

CHAPTER 9

Working with Contract Research Organizations

WHY BUY IN R&D SERVICES?

A company that exists to provide services to other organizations and whose 'products' are knowledge-based, the fruits of scientific and/or technological research, is generally known as a contract research organization (or CRO). Very often they will serve a niche sector, as the following UK examples illustrate:

- Paint Research Association
- Electrical Research Association
- The Welding Institute
- Motor Industry Research Association
- The Forensic Science Service (law enforcement)
- AEA Technology (nuclear industry)
- Nirex (radioactive waste industry)
- Scott Polar Research Institute
- Building Research Establishment.

Other organizations serve a broader range of customers, but always with the accent on knowledge-based, research-intensive services. Again examples from the UK:

- Roke Manor Research
- QinetiQ
- Marconi Caswell Technology
- The Technology Partnership plc
- Quintiles Research
- Cambridge Drug Discovery Ltd

- AEA Technology
- The Automation Partnership
- The Babraham Institute (biosciences).

There are also company-specific in-house, or otherwise 'captive' research and development units, that will sometimes sell their services to external clients, where this is commercially convenient to them. All these organizations, then, are CROs in the broadest sense. The subject of contracting with them is covered exhaustively in Gower's *The Outsourcing R&D Toolkit.*[1] The purpose of this chapter is to locate the use of CROs in the context of knowledge buying strategies, to assess where they might add value, and to suggest mutually beneficial ways of working with them.

One key area of knowledge creation, as has been suggested throughout this book, is the area of technological and scientific *research and development.* This is becoming a major subset of knowledge procurement, and not only for commercial organizations. Across most enterprises, State or private, managers are expected to use the best and most up-to-date knowledge to steer their organizations and maximize value delivered through day-to-day operations. Where knowledge does not yet exist, there is a powerful drive to create the knowledge, if necessary from first principles. At the beginning of the twenty-first century, globalization has forced organizations – especially private companies – to become more effective at developing new products, processes and services as increasingly fierce competition threatens virtually all markets. Technology is becoming ever more complex, product life cycles are shortening and substitute technologies follow in rapid succession. This is true of every industry and every activity that has a technology/knowledge basis. Companies find it increasingly difficult to carry the internal resources to sustain R&D capability across all disciplines and so have for the past 20 years focused internal research on core competencies whilst outsourcing (in the broadest sense) non-core research by forming strategic partnerships with other organizations or contracting out the research of non-core activity. The future will undoubtedly see greater contracting-out of core research as well.

It has already been noted in Chapter 5 that purchasing R&D services is a form of temporary integration of the buyer and the seller. Contract research generally implies a higher level of integration, as the research contractor is bound by terms and conditions to deliver certain results/information (knowledge). The research contract may bear some similarity to an employment contract, for example in terms of exclusivity and confidentiality which should outlast the contract itself. There need to be clear boundaries in the minds of both buying and selling organizations, especially

1 See Appendix 1 for full list of contents.

where individuals will be working very closely with their opposites, to avoid the danger of unduly cosy relationships being established, possibly to the detriment of forming other useful, competing, relationships. Whilst there is a good measure of truth in the quote below, it also unintentionally illustrates a danger of two organizations becoming too closely aligned:

> The linkage established by contract research substantially differs from sub-contracted production, or services such as maintenance or routine analytical work. In many ways this linkage is a partnership akin to a car manufacturer hiring a design studio for the conception of a new model. It is a bilateral, formal linkage, limited in time, targeted at a specific objective, and requiring, prior and during the collaboration, considerable intimacy and mutual understanding between the two partners with regard to expectations, goals and capabilities; for the duration of the project the contractor constitutes a true extension of the client's R&D organization.[2]

Typical reasons for using external R&D contractors, as identified in a study by the world's largest CRO (Battelle, Geneva) are, in descending order of precedence:

- *Innovation* – client seeks competitive advantage through product or process improvement or acquiring technology with which it has limited experience. Environmental legislation is an increasing spur to innovation. In these cases the client does not necessarily define the solution, but may be interested in having a solution 'sold' to them.
- *Problem solving* – client encounters unexpected problem and needs agile 'off the shelf' expertise. The CRO is then a 'firefighter' valued for its experience, expertise and ability to respect confidentiality.
- *Additional capacity* – client has expertise but other priorities prevent internal resource allocation. Work is outsourced for a limited time to complement in-house work. This may also reflect an inevitable consequence of downsizing where (perhaps) strategic managers lack the vision, confidence or technical grasp to manage in-house resources and see contract research as a way of hedging risk by transferring some of it outside.
- *Multidisciplinary* – client finds it difficult to mobilize a team in-house with the necessary range of technical knowledge/expertise. A CRO may be able to integrate the necessary expertise in a flexible way, bringing together a team under one roof.

2 'Stretching the knowledge base of the enterprise through contract research', International Institute for Management Development, Lausanne, Switzerland. *R&D Management*, vol. 22 no. 2, April 1992, p. 178.

- *Cost sharing* – clients, perhaps in the same industry, working temporarily with competitors, or across industries with a common interest, identify work that can usefully be shared. This is often high-cost and/or high-risk. Cost-sharing becomes an attractive strategy and also allows diversity of project management, perhaps enhancing the value of the results.
- *Business intelligence* – by hiring a CRO a client can gain access more quickly to a wider range of intelligence or focus the resources necessary for effective intelligence acquisition. The CRO, as an 'expert' in its own right, has access to and perspective on available information and is well placed to locate and evaluate any necessary new information. In a commercial setting, an effective confidentiality clause will be necessary to prevent the client's competitors acquiring the same intelligence.

As part of the 'due diligence' process discussed in Chapter 6, the CRO must convince its potential client that it will be the best partner to work with. A considerable amount of problem definition will take place during the negotiation process which will normally cover both business/contractual and technical matters. The client organization's principal negotiator will usually be the engineer/scientist who will have day-to-day contact with the CRO. The negotiator must be convinced by the capabilities of the CRO and supplied with sufficient arguments (knowledge) so that in turn he or she can convince their manager of the value of the proposed contract. These negotiations will be protracted, typically a period of six to nine months from first contact to contract placement.

STRATEGY ISSUES

As late as the mid-1980s, the principal buyers of R&D services were larger companies and organizations with in-house R&D departments – typically in sectors such as government, energy, chemicals, pharmaceuticals, electronics, IT and transport. The reasons for this were that these 'buying' organizations needed a good understanding of risk, timescales and technology potential. Many of these organizations work on long-term horizons, with current investments having multi-decade paybacks, often within evolving regulatory environments that also require long-term research (for example the electricity industry's research into the effects of atmospheric pollution). But in the last few years of the twentieth century the need to buy in R&D became an increasingly common fact of life for smaller organizations. High-tech companies interact with mul-tech clients and must themselves be able to cover a broader spectrum of technologies, buying in those that are complex or costly, or which are too far removed from core business activity.

'If we don't sell, someone else will!' The age-old justification for dubious arms sales is also a stimulus for external R&D. Competitive supply of new technologies, especially where buyer and seller are not direct competitors, and where there are often many competing suppliers, increases the pressure to supply R&D services. If we do not sell, a competitor may capitalize on some competing technology which may in turn permanently distort our particular market. A classic example of this was the competitive battle between VHS and Betamax technologies in the video recorder market, which VHS won so comprehensively that Betamax was rendered extinct in the space of a few years.

A corollary is also true: if we don't buy, someone else will! There is a legitimate business case, for firms that can afford to do so, to acquire the services of external suppliers of research and development services to deny their competitors the opportunity to use them. In the UK traditional vacuum cleaner manufacturers have candidly regretted that they turned down the opportunity to acquire and shelve the technology associated with the revolutionary Dyson bagless vaccum cleaners that today have rendered 'bag' vacuum cleaners all but obsolete. There are fewer public cases of research buyers awarding small and ongoing research contracts to inventors and technology firms to prevent them from developing new relationships with competing firms. If the inventor is a one-man band he or she can only work for one customer at a time; the technology firm may be unable to develop new clients due to their ongoing exclusivity arrangement with the research buyer.

The Ove Granstrand report noted a distinction in the commercial sector between 'high-tech' and 'mul-tech' organizations:

> Technology diversification at product level justifies the notion of multi-technology products – or 'mul-tech' products for short. While 'high-tech' products typically refer to products with a high R&D content (high R&D intensity or high R&D value-added) based on recent technological advances not widely available on technology markets, 'mul-tech' products refer to products that are based on several technologies, which do not necessarily have to be new to the world or difficult to acquire. (The two concepts are not mutually exclusive, however.)[3]

Individual products can feature both high-tech and mul-tech characteristics, but the management of the development of these products will differ in emphasis and style. The *technology management* of high-tech products focuses on:

3 'External technology acquisition in large multi-technology corporations', O. Granstrand, E. Bohlin, C. Oskarsson and N. Sjoberg, Department of Industrial Management and Economics, Chalmers University of Technology, Gotenborg, Sweden. *R&D Management*, vol. 22, no. 2, April 1992, p. 127.

- scientific creativity
- advanced customer demands
- elitist recruitment
- technological leadership
- technology protection.

The *technology management* of mul-tech products, by contrast, focuses on:

- technology scanning
- incremental technology improvement
- imitation strategy.

The difference between high-tech and mul-tech is analogous to the difference between '*R*' (for research) on the one hand and '*D*' (for development) on the other. To manage technology and innovation effectively in the present mul-tech environment demands a *broader range* of management skills (social, economic, legal, engineering, outsourcing) than traditional inhouse R&D management. Mul-tech and high-tech management must play the technological market effectively. External technology acquisition is complementary to, and not a replacement for, in-house R&D. In-house R&D is often needed to absorb/internalize external technology.

Is there an optimum degree of external technology acquisition? This is analogous to the whole question of outsourcing non-core business activity and there is already a great deal of literature to alert senior managers to the broad issues. With regard to technology or knowledge 'outsourcing', the danger of relying on external sources is a real one: what happens if the external source ceases to be available? In the real world however, rising R&D cost and risk is stimulating greater demand for external technology that is being matched by an increased supply of sources, especially niche suppliers of special technologies that may have numerous applications.

The extent of external technology acquisition is a strategic choice for each business – whether State or private, industrial, commercial or government. Total external technology acquisition (outsourcing in the purest sense of the term) is an unlikely option, although a few brave commercial firms are beginning to experiment with the idea. But why is total outsourcing unlikely? Having something to offer potential suppliers/partners based on unique in-house R&D capabilities in complementary technologies provides a 'ticket of admission' to patent pools, cross-licensing arrangements, collaborative research and development and so on, whilst also offering the most secure route to internalize bought-in knowledge and technology. Some commentators suggest that businesses will increasingly need to *compete* to be viewed as prestige/quality *customers* of their various supply markets. This remains to be proved, but in the technology field, being a desirable client, in the sense of providing good research 'suppliers' with useful, leading-edge work to do, will always be more attractive to them than taking on assignments that involve the research supplier (in effect) carrying an ignorant client, even if, over the short term, it may prove financially more rewarding!

CONTRACT RESEARCH AS A STRATEGY TO INCREASE INNOVATION

Innovations may be defined as products and/or techniques of a new quality that an innovator introduces to the market for the first time. Innovations may occur at many stages in the development or life-cycle process from discovery through development through diffusion. Innovation processes are primarily *information processes* in which knowledge is acquired, processed and transformed. Whilst information flows accompany even routine transactions, such as selling, tax raising and regulation, where innovation is the prime goal, the acquisition, transfer and processing of information is managed proactively in such a way as to maximize the usefulness of the information for the innovative task.

Research client organizations need to recognize the information system *as a system* where there are senders and recipients of information. As a rule, information relationships tend to be asymmetrical in the sense that in the initial phase, the sender has a qualitatively higher level of knowledge than the recipient – the information exchange process is aimed at reducing this asymmetry in a managed way, so that eventually knowledge is equalized. Knowledge equalization is an important objective of most research contracts!

One problem that the innovator must overcome is that, often, information supply is too great. It has been noted in studies on innovation management that *information search* (as opposed to information supply) is a critical success factor. This has implications for researchers in determining their research strategy and, where that involves buying in research services, implications for the way in which the research specification is drawn up. Information supply can lead to information overload – a noted cause of failure – and one that will continue to accelerate as a cause of failure in today's information explosion. Information search implies discrimination in the use of available information, as Internet search engine suppliers are discovering to their benefit! Jurgen Hauschildt, in his study 'External acquisition of knowledge for innovations',[4] commented on this subject of information supply:

4 'External acquisition of knowledge for innovations – a research agenda', Institut fur Betriebswirtschaftliche Innovations Forschung der Christian-Albrechts-Universitat zu Kiel, *R&D Management*, vol. 22, no. 2, April 1992, p. 107.

> Coping with information in innovative situations is a matter of developing appropriate patterns of connecting different items of knowledge. Once the decision maker is offered a certain scheme for this connection he immediately and dramatically increases his performance. Thus, information supply will not be successful where it merely delivers data. It can be effective only when it presents the connecting patterns or when it stimulates the information user to develop them on his own.

Hauschildt quotes Gemunden's earlier study, which noted that the level of aspiration of the acquirer of knowledge is an important factor. By aspiration he means the extent to which the buyer aspires to acquire a thorough understanding of the knowledge being transferred. The lesson has obvious application across the field of knowledge-transfer relationships:

> If a solution with a low level of aspiration is intended, the seller organization should dominate the interaction and the decision process should focus on the technological aspects of the innovative problem. If a solution with a high level of aspiration is intended, both parties should be involved into an intensive interaction process during which technological and organizational problems of innovation use are analysed ... The interaction of the seller and the buyer of innovative products is successful when each party is willing and able to familiarize itself with the domain of the partner: the seller must learn about the application needs of the buyer and the buyer must work his way into the technological details of the seller.[5]

The lesson for the buyer of R&D services is obvious, but by anecdotal evidence seems frequently to be missed. Make sure the buying organization is clear about the level of knowledge transfer required – in other words be clear about your aspiration – and make sure this is clearly reflected in the work specification. During contract progress meetings, the research client organization's technical representative must ensure that the correct level of knowledge transfer is being achieved.

5 H. G. Gemunden, 'Effiziente Interaktionsstrategien im Investionsgutermarketing', *Marketing ZFP*, 1980.

WHY BUY KNOWLEDGE WHEN YOU CAN BUY THE OWNER OF THE KNOWLEDGE?

This is restricted to the business realm where buying a knowledge-rich firm by merger or acquisition may be a viable proposition. Suverkrup[6] has studied the factors associated with successful technologically-oriented corporate acquisitions which he found to be dependent mainly on the quality of the post-acquisition integration process, based on efficient decision-making and the support of experts. Conversely, unsuccessful technology acquisitions were those where highly formalized integration/absorption takes place, treating the target firm as an exploitable knowledge-mine and reducing its autonomy.

The most successful technology-based motives for acquiring another firm are:

- exchange of know-how
- high-productivity R&D
- easy recruitment of researchers
- completion of a particular product programme.

Where these motives are not present, the acquisition of a firm for technological reasons may reflect a short-term tactical move rather than a strategy, unless of course, the strategy is simply to take over a competitor to reduce competition! Merger and acquisitions is a subject outside the narrow scope of this book, being already well served by a considerable body of literature. The main point to be made here is that buying a company as a short-term expedient solely to acquire its R&D capabilities is virtually unknown. A company acquisition must in any case be considered in the context of competition law. The decision is likely to depend on

- the acquirer's overall strategy
- the target firm's overall health – technical, financial, managerial, patent portfolio
- the general state of the market
- what competitors are doing.

Another method to 'buy' the owners of knowledge is the recruitment of key staff as permanent employees, possibly using head-hunters to target the right individuals.

6 C. Suverkrup, 'Ziele und Erfolg internationalen-technologischen Wissenstransfers durch Unternehmens – akquisition: Eine empirisch Untersuchung am Beispiel deutschamerikanischer und amerikanisch-deutscher Akquisitionen,' *Diss*, Kiel, 1991.

Mighty software company Microsoft has taken this concept about as far as any company ever has. A fascinating insight into Microsoft's approach to R&D was given in a report in the UK *Financial Times*[7] at the time this book was being prepared. The management philosophy at Microsoft was to recruit widely and allow their researchers to set their own agenda, their employer merely providing an environment in which they can shine. 'I see myself as a ringmaster' said Andrew Herbert, the director at Microsoft's Cambridge UK laboratory. 'Here are a collection of great performers. My job is to make sure they can put on the best show they possibly can.' For a scientific researcher this sounds like research-paradise. One of Microsoft's top scientists spent the first few years of the twenty-first century linking the world's astronomy databases – not a feature that obviously lends itself to be featured in the next release of the Windows operating system, the staple product of the company.

Microsoft's researchers, being amongst the world's best ('brilliant' in the words of the *FT*) were free to advance the state-of-the-art in software and computer science in areas of their choosing. But Professor Hank Chesborough (executive director of the Center for Technology Strategy and Management at the Haas School of Business in California, was quoted by the *FT* as a sceptic:

> When you bring in so many really talented people there is a natural tendency for them to justify themselves by their own ideas and inventions, not by their ability to spot really important ideas. There is a marginal tendency to talk down what is available on the outside.

Technology consultant Georges Haour, professor of technology management at IMD, also sees dangers in depending too much on internal knowledge generation: 'Technology firms must break away from internal innovation and redefine their *innovation perimeter* to include ideas that come from outside', he said. A strong argument for maintaining a diversity of knowledge creation strategies, including judicial use of contract research.

The third method of acquiring the 'owner' of knowledge, or at least privileged access to that knowledge, is via technology licensing. Licensing is a form of technology transfer in which the *licensee* acquires a right to use a technology from the owner of the technology (the *licensor*). Technology can, of course, be transferred simply by purchasing a particular machine or piece of equipment, but this can also be done by others – for example, competitors. The need for a licence arises where access to the technology is restricted by a patent or other IPR protection and it becomes necessary to enter into a legally binding agreement with the IPR owner in order to acquire the right to use the technology.

---

7 S. London, 'Good old-fashioned innovation', *FT*, 12 March 2004.

The licensor usually grants a licence in exchange for a fee on signature of the licence agreement and subject to a periodic royalty which represents an agreed percentage of the utilization of the licensed technology – normally in terms of products sold during the life of the agreement. In order to earn a suitable return from the licence the licensor will seek assurances from the licensee as to how the technology will be utilized and will satisfy itself that the licensee is, overall, a reliable partner. This may be underwritten by a minimum royalty clause, to focus the licensee's attention. In granting a licence to a particular licensee, a licensor often has to forgo licensing to other parties, especially in situations where an exclusive licence has been agreed, so it is only common sense to ensure that the licensee is a serious business partner.

Licensing may be an attractive alternative to R&D work: R&D is expensive, carries heavy fixed costs in terms of personnel and infrastructure, and takes time to yield benefits. Some commentators have suggested that R&D is ten times more expensive than licensing as a strategy for acquiring new technology, and a number of organizations have made a conscious choice towards licensing as the basis of their knowledge acquisition strategy. Licensing, therefore, should be considered as a strategy in the overall project selection process. It should be noted, however, that licensing has some disadvantages:

- It ties the licensee to a particular technology for a long period.
- It may damage the licensee's long-term technological competence. As noted elsewhere, the ability of a company to bring 'something to the party' in terms of R&D often serves as a means of gaining privileged access to leading-edge technology with all that this implies in terms of competitiveness.
- The loss of internal R&D capability inevitably leads to a reduction in the organization's ability to comprehend and internalize new technologies as they arise – and some lost skills are impossible to replace, no matter how much money is thrown at the problem!

HOW TO WORK SUCCESSFULLY WITH A CRO

A number of critical success factors need to be present and operating for there to be any real prospect of external R&D project success:

ADEQUATE PROJECT APPRAISAL

Without an adequate programme/project appraisal process, examining:

- the characteristics of the client organization (strengths and weaknesses),

- its knowledge absorbative capacity
- the market for any resulting technology (or use to which new knowledge will be put)
- operating environment in which the knowledge is required
- the relative advantages of in-house R&D over contract R&D

the client organization will be unsure as to what are its overall objectives. Some of this work, of course, is routinely undertaken as the organization sets out its multi-year strategic plans and builds its annual budget forecasts. If the client organization is unclear about what the ultimate requirement is, it will find it difficult to guide the CRO.

Any organization, commercial, governmental, or non-profit, that has a requirement to invest money to acquire new knowledge or technology, will have a process through which it decides (a) what are its objectives (b) on what it will spend its money and (c) the method of monitoring and final evaluation of such projects. This is called a project appraisal process. The appraisal process will encompass the following steps that ideally should be carried out sequentially, as shown in Table 11.

Table 11 Project appraisal – steps

Organizational objectives	*Project objectives*
Technology programme appraisal = programme justification and execution plan	Project appraisal = project justification and execution plan
Technology programme monitoring = programme history and records	Project monitoring = project history and records
Technology programme evaluation = consolidated programme history and reports	Project evaluation = consolidated project history and reports

ADEQUATE SUPPLIER ASSESSMENT

We have already looked at supplier assessment in Chapter 8. Inadequate supplier assessment represents a risk that is difficult to quantify. Good quality research supplier assessment represents a considerable opportunity to award work to the most suitable CRO and to develop new 'supply' sources that increase the depth and breadth of the client organization's technology/knowledge portfolio.

ADEQUATE WORK DEFINITION

Again, we have already looked at client technical specifications in Chapters 6 and 8. External work must be adequately defined in terms of:

- inputs from the client organization as buyer
- inputs from the contract research organization
- inputs from third parties (and who is responsible therefore among the contracting parties)
- additional help to be provided by the client organization during the work
- outputs required – When? Where? What?

These questions are the practical ones that follow on naturally from the important strategic ones of:

- Why do we fund R&D on this occasion?
- What do we want from the R&D?
- What would be the effect if we did nothing?
- What will we do with the research results/technology?

A 'TEAM' PHILOSOPHY

In R&D contracting there tends to be a natural affinity between professionals working in the same field, pushing at the same knowledge frontiers and often sharing similar academic and work–life experiences. To that extent the prevailing ambience tends to be collaborative and cooperative. This does not mean, however, that effort should not be made by both parties to develop and foster a team approach to the project in hand, providing that this approach does not become one that either party exploits to its own narrow advantage.

How can a team philosophy be encouraged? There are a number of strategies based on normal team-building techniques and identifying shared interests. We will concentrate briefly on the contractual document that underpins the relationship – an area normally overlooked by project managers, if not their lawyers. The form of contract should support the project management method adopted by the parties. Ideally it should not engender an adversarial attitude. It is arguable that the best format of commercial contract available in the world today is the New Engineering and Construction Contract (commonly called the NEC) published in the UK by the Institution of Civil Engineers. This observation is made because the NEC actively seeks to be a project management tool and to engender a team approach and philosophy to project management in an industry notorious for its confrontational and litigious style. At the time of writing the NEC, having been developed as a family of contracts with a specific consultancy sub-form (the NEC Professional Services Contract) it appears to be having remarkable success in the engineering sector and is now being applied more widely. Certainly it represents the biggest development in contracting practice for 100 years. On major projects using the NEC, there is normally a project launch meeting where the various parties meet and determine their immediate priorities. Some have used these meetings to team-build in a proactive manner and supplement them with regular briefings with team-building primarily in mind. Whether this is appropriate in an R&D project will depend very much on the particular circumstances and to the overall size of the project.

CONCLUSION

Among the various strategies for buying in new knowledge, the use of contract research organizations represents a well-proved mechanism. With many major independent research organizations now fifty or more years old, and with a global R&D marketplace opening, it is certain that many managers will turn to these organizations to assist them to meet both short-term tactical needs and long-term strategic goals. The main practical obstacle to using them may be in the area of intellectual property rights. Enlightened CROs and client organizations seek to be realistic about such rights. User rights are often more important than ownership rights.

CHAPTER 10

Knowledge Factories – Buying Knowledge from Universities

UNIVERSITIES AS KNOWLEDGE BROKERS

The teaching of students is the primary objective of universities and, even allowing that their students represent the bulk of the future intellectual capital of the nation, they are by definition only partly trained and inexperienced. A university may not at first sight appear to be the best place in which to undertake research or other knowledge-procurement activity. Yet universities are widely seen as being not just a creator of knowledge and trainer of young minds, but increasingly as a major agent of economic growth. *The Economist* newspaper, in a major survey on universities, adopted the phrase 'the knowledge factory' to describe the way in which governments and industry increasingly perceive these institutions as:

> a major agent of economic growth: the knowledge factory, as it were, at the centre of the economy. In such an economy – one in which ideas and the ability to manipulate them, count for more than the traditional *factors of production* – the University has come to look like an increasingly useful asset. It is not only the nation's R&D laboratory, but also the mechanism through which a country augments its 'human capital', the better to compete in the global economy.[1]

For the purposes of this chapter, we refer to universities and other institutes of higher education simply as 'universities'. From the point of view of the knowledge-buying client organization there are a number of different types of potential knowledge-contractor within the tertiary education system and they are not all universities. However, in practice the bulk of sponsored research/knowledge creation work performed in institutes of higher education is performed by universities. In most of the developed countries a large share of research and development activity, typically 15%, is performed in universities. The real significance of the universities' contribution is greater, however, since they conduct the bulk of basic research. The majority of university research is funded by the government and the public sector. The general belief is that without the State's involvement in basic research, industry would not make up the shortfall because companies cannot find a way of securing the benefits of

---

1 'The knowledge factory – a survey of universities', *The Economist*, 4 October 1997, p. 4.

such research exclusively for themselves. Governments also accept the 'received wisdom' that basic research contributes directly to economic growth.

There are however critics of the flow of research money into universities. Upholders of the *academic tradition* of universities argue that external research funding undermines the independence of universities, increases the dominance of research over teaching, promotes some disciplines unfairly at the expense of others and represents a policy by the State biased towards the creation of intellectual property. The consequences of this to scholarship, they argue, may be as damaging as State-driven industrial policies have so often been to economic development. Governments for their part predominantly see basic science as a 'public good' and have, over the past two decades, sought to spare university science from the cutbacks they have inflicted on their own laboratories. Countries such as Japan and France, where universities have traditionally held a smaller role in research, plan to increase their funding in this area – although at the time of writing both countries face severe budgetary constraints which may undermine this general intention. The intention to maintain and increase university research funding is partly because of the success of some universities in the *added value* of their R&D activities. A study by Bank Boston in the late 1990s concluded that if the 4000 companies funded by MIT graduates and faculty were turned into an independent nation, the income they would produce would make it the 24th richest in the world.

With the end of the Cold War, which was an important stimulus to government investment in R&D generally, global economic competition has become a fresh justification for continuing high government investment in basic science. In the US it is noted with some concern that companies such as General Electric and AT&T, which used to undertake all their own basic research (sometimes earning Nobel Prizes on the strength of it) are now concentrating most R&D effort in product development. Governments are interested as never before in the way their funds are invested by universities which have traditionally received funding via block grants to be divided between teaching and research at the discretion of the university. OECD analysts, however, have detected a growing share of mission-oriented funding in recent years, accounting in the UK for 30% and in the Netherlands, 15% of the total, which channels funding to the support of specific technologies. This may be good news for knowledge-hungry client organizations, where the favoured sciences happen to be in their field of interest.

Universities have proved an enduringly successful beneficiary of R&D investment. As stated 35 years ago in the *Handbook of Industrial Research Management* (Reinhold Book Corp, 1968, page 94): 'Wherever people are being trained to do research, they must be doing research; no substitute for this approach has ever been found. Therefore every such institution is a centre of research activity.' They are endowed often with world-class facilities, world-renowned academics and gifted and enthusiastic undergraduates. Small wonder, then, that, despite the tension between teaching and research, they prove to be a fruitful R&D vendor for research client organizations.

PROBLEMS IN BUYING RESEARCH FROM UNIVERSITIES

In planning to buy knowledge-based services from universities, the knowledge-buying client organization must undertake all the normal procurement-type due diligence suggested earlier in this book. There are, however, special problems in dealing with universities. The UK's Centre for Exploitation of Science and Technology (CEST) conducted a survey in the early 1990s of companies that had used universities for contract research. Of those surveyed 70% had used universities as a limited source of technology and, surprisingly, 40% had used foreign universities. Their experiences were varied, however, as these quotations reveal:

> *The trouble is, they can't come up with the goods on time.*
> Chief Executive, Oil and Chemicals Industry

> *We took a policy decision some time ago to direct 10% of the R&D budget (£1m) to UK universities – we wrote to 120 universities (46 of them twice) only 30 replied, 12 said they couldn't visit us, 12 others came and half of them left, uninterested! Later we joined the MIT research programme: it's very good, a model of how it should be done in the UK.*
> Corporate Development Director, Motor Vehicles and Spares Industry

> *They're very good value for money. But you need to make them feel involved, you need to find the ones who want to make money – that's why the squeeze on funding is good news.*
> Chief Executive, Electronic Equipment Industry

> *We would only use universities where we can quite specifically define the problem, the sort of solution and the performance we want to achieve.*
> Managing Director, Aerospace Industry

> *An under-utilized resource. They need to be used in the right way (hands-on, precise project definition, use them to produce a device rather than just some data/reports, encourage an interdisciplinary approach – it's mutually more satisfying). The cutbacks in funding are good because it encourages industry/university interaction – although it mustn't go as far as jeopardizing basic science.*
> Technical Director, Oil and Chemicals Industry

The CEST survey concluded:

> The consensus is that universities can represent a valuable and cost-effective source of science and technology provided the relationship is managed properly and the outputs are precisely specified. ... It is important that both parties understand the constraints on the other and are clear about the expected deliverables. In practice this means substantial management time must be devoted to ensuring the success of the relationship, though the cost should be amply repaid. universities provide a valuable source of knowledge at the forefront of technology and they can fill a skills gap in the company. Maintaining a high profile with a university also helps with the recruitment of its graduates.

Universities are very focused upon the opportunities for commercial exploitation of the work they are involved in, but often have unrealistic intellectual property (IP) objectives. Much valuable management time is wasted negotiating IP considerations on relatively low-value contracts where the likelihood of direct commercial exploitation is remote. Traditionally, investment in university research has been considered by business to be at least partly investment in the national fabric (or, as is popular in the UK, investment in 'Great Britain plc'). Universities have gratefully seized on such investment as a way of subsidising teaching.

The balance, as we have seen, is shifting. When asked whether their interests lay primarily in teaching or research, respondents to a 1996 survey carried out by the Carnegie Foundation for the Advancement of Teaching, across the universities of a number of countries, found that there is often a preference towards research (see Table 12).

Table 12 Universities: loyalties to teaching versus research

	% primarily in teaching	*% leaning to teaching*	*% leaning to research*	*% primarily in research*
Australia	13	35	43	9
United Kingdom	12	32	40	15
Germany	8	27	47	19
Japan	4	24	55	17
Russia	18	50	29	3
Sweden	12	21	44	23
United States	27	36	30	7

In the UK, before publication of the Cooper Report in 1989 – a report which has been influential in other countries with large tertiary education sectors – universities looked primarily for marginal recovery of overheads (usually, in the UK, at 40% of staff costs). Today a much higher rate is often sought. Knowledge-buying client organizations should note that overheads should generally be paid only on staff costs, as the costs of university contract research are primarily people costs. The exception to this is where the university invests in capital equipment with specific reference to a prospective contract from a commercial client. If the client is not directly funding that investment under the contract, then overhead recovery needs to take this investment into account. In these cases, as the university has borne investment risk, it is proper that some overhead recovery should be payable on this investment.

A price build on a university research contract which is expected to continue for several years may include the following headings:

- Staff costs
  - Research assistant
  - Research student
  - Research supervisor
- Superannuation/NHI
- Overheads – at [x] per cent of staff cost
- Equipment – payable at cost
- Consumables – payable at cost
- Travel and subsistence – payable at cost at standard academic rates.

In the UK context, VAT may need to be itemized separately. In other countries, local purchase or value-added taxes may be payable. Advice should be sought from the university in the early stages of the negotiations.

In a research project lasting a number of years, fixed prices are unlikely to be appropriate because academic rates will almost certainly be inflated year on year: to meet these increases will probably be less expensive than meeting the university contingency if fixed prices are sought. A maximum estimated price must then be agreed. Prudently the knowledge-buying client organization will build in an undisclosed contingency figure but it should not be assumed that this must be spent – it is a contingency.

A cost control clause in the contract will be necessary. The following is typical:

- Salary costs are subject to increase arising from incremental and nationally agreed salary awards only, and to prior notification by the university to the purchaser in writing.

- None of the itemized sums in the price breakdown shall be exceeded without the prior written approval of the purchaser. An underspend on one of the itemized sums may not be used to fund an overspend on another itemized sum without the purchaser's prior written approval.
- If and when the cost of the service, including commitments, amounts to 80% of any of the annual/total/itemized sums the university shall at once inform the purchaser in writing stating his or her liability to date and the estimated cost of the work remaining to be done.
- The university shall advise the purchaser forthwith in writing if at any time it becomes apparent that any of the annual/total/itemized sums will be exceeded. The purchaser shall then decide whether or not the service shall continue.
- The purchaser shall not be liable to meet any costs incurred or committed in excess of the agreed sums or if the service has not been carried out in accordance with the contract.

It is useful to ascertain the proportion of time to be spent by each grade of research staff on the project in question. Supervision, in particular, is likely to be a small proportion of the supervisor's total time. Caveat emptor!

PITFALLS TO BE AVOIDED

There are a number of common pitfalls to be avoided when dealing with universities on research contracts:

- *The research student may not complete his/her course and may therefore not complete the research project.* The pre-contract interview and selection process may reduce this possibility, giving both the university and the client organization the opportunity to probe the student's commitment and staying power, but the risk cannot be ruled out altogether. A financial retention may be useful – certainly if the final report is not delivered then the final payment should not be made to the university and, depending on the terms of the contract, it may be appropriate to recover some or all sums already expended. If, however, a draft report has been received, failure to deliver the final report may not be deemed to be an insuperable problem to the client organization.
- *Progress reports may not be supplied without prompting.* Universities have

traditionally been less professional than commercial CROs in project management. This shortcoming is improving year on year, particularly as universities become increasingly focused on external research work. The importance of good, proactive project management on the part of the client organization cannot be overemphasized, however (see Chapter 6). The client organization should not be satisfied with anything less than good project management from the university. This is after all, one reason why he or she often pays separately for supervision!

- *The university will wish to publish results, particularly if the research project is part of the student's thesis (which often must be made public under the university's Articles of Incorporation)*. In many situations it will not be a problem for the client organization to agree to publication: but it obviously depends on the nature of the work and the need for confidentiality. The most difficult problem may occur in situations where unexpectedly valuable results emerge, or results that may prove to be a severe embarrassment to the research buyer and must be suppressed. The conditions of contract should include a provision that the client reserves the right to veto publication. Expect this to be a subject for negotiation. It is possible, subject to the agreement of the university authorities, to have theses published on a need-to-know basis. This, in effect, debars the university's information service (or library) from making the thesis generally available for a predetermined number of years.

- *Staff may change over a long project.* The client organization should stipulate that staff changes (or at least, approval of replacement staff) are subject to its prior written consent.

- *The university may require a long period following notice of termination.* Termination for convenience is not really in the spirit of research contracting with universities. If the work is proving to be unsuccessful, both parties normally attempt to renegotiate the contract to see if the project can be reconfigured to increase likelihood of success. A full academic term's notice of termination is not unreasonable, given that the university will have a responsibility to the student to redeploy them onto another project meeting their educational needs and aspirations. What 'a full academic term's notice' means in practice to the client organization is that there is a termination fee (except, of course, in cases of default or breach) which is equivalent to the student's cost for a full academic term.

- *The university may have poor cost control methods and expect to bill the buyer months or even years after the event!* As noted above, universities are getting steadily better in the UK at professionally project managing their external research contracts. However, a clause in the contract to limit the university's right to submit bills long after the event is a useful device for defeating this problem. A clause along the following lines may suffice:

> The [client organization] shall not be liable to meet price increases where the university fails to notify it in accordance with [relevant sub-clauses] or in any other circumstance where notification of or application for price increases is unreasonably delayed by the university. For the purposes of this clause a reasonable period shall be deemed to be one academic term from the effective date of the price increase.
>
> The [client organization] shall not be liable to pay any invoice where the university fails render a correct invoice in a reasonable time. For the purposes of this clause a reasonable period shall be deemed to be two academic terms from the date the invoice should have been so rendered. Time of rendering of invoices shall be considered to be a condition of this contract.

RESEARCH FELLOWSHIPS

Knowledge-buying client organizations may wish to link with universities in other ways. A Research Fellowship is a prestigious university appointment, normally at post-doctoral level. Appointments are usually for three years but may be extended. The terms of reference for the fellowship are negotiated individually but there is an increasing tendency to require the Research Fellow to be closely involved with some major topic of interest to the client organization. The Research Fellow is normally paid on the same scale as a university lecturer of similar age and experience. Academics may also be retained on consultancy contracts or short-term employment contracts – specific terms of reference will be necessary.

ROYALTIES

The payment, or not, of royalties has become a battleground between universities and their industrial/commercial partners in the UK in recent years. This was encouraged by the Cooper Report of 1989 which reinforced the pre-existing argument of universities that they were subsidizing industry research. The Cooper Report stated that

> For the purposes of this report, we have considered only ventures involving an HEI (Higher Education Institute) in research for which partial funding is received from industry, the HEI putting resources in

> e.g. in the form of knowledge and experience of researchers and the associated costs of developing specialised groups[2]

and went on to state that it had excluded from consideration 'fully funded research contracts at universities'. However, the report failed to define what was meant by 'fully funded'.

The Cooper Report suggested that in the UK the most common contractual arrangement in research contracts with universities was for 'the sponsor to own the IPR but to pay the HEI reasonable royalties in recognition of their contribution to the project' (p. 7). This rather sweeping statement may, in its reference to royalties, have been more akin to a wish than to reality. Certainly many industrial sponsors do not consider it appropriate to pay royalties on work they have had undertaken in universities in exactly the same way that they do not expect to pay royalties to independent contract research organizations.

Where the university has first approached industry with background IPRs and with a concept for industry to take forward, develop and commercialize, there are good grounds for ensuring that work is fully recompensed via a royalty arrangement. Where, however, industry is proactive in searching out sources within universities to undertake R&D work, is investing its own energies, resources, people and commitment to enable R&D projects to be taken forward, not to mention their contribution to the education of under- or postgraduates, the argument for royalties is much harder to sustain. It is likely that the Cooper Report in the UK was widely misunderstood and misapplied by the university sector. The arguments against paying royalties for university contract R&D work revolve around the idea that for multi-party, multi-year, multitechnology projects it is impossible to determine the value of the various contributors to the project.

Even in less complex technological developments which eventually reach the market, it can take the wisdom of Solomon to determine the value of an university's 'contribution' to a project's commercial success.

The university argument that they subsidize the R&D through the knowledge and experience of researchers and the associated costs of developing specialized groups, seems to be implausible. First, it is very unclear that the costs of such know-how are not already recovered in the overhead rate applied to staff and other costs. Second, and of more relevance, when placing a contract with an independent contract research organization (CRO), or a consultant, or indeed a manufacturing firm for

---

2 *Intellectual Property Rights in Collaborative R&D Ventures with Higher Education Institutes*, UK Department of Trade and Industry Interdepartmental Intellectual Property Group, September 1989, p. 3.

services or goods, one does not expect to be charged an additional success fee on the basis that the CRO or manufacturer has 'knowledge and experience' which has contributed to the project. These costs should rightly be reimbursed in commercial overheads.

Where, for whatever reason, royalties are payable on the exploitation of the research results, they are defined on principles similar to those found in licensing agreements, that is:

- the royalty base – the precise definition of the products in respect of which royalties shall be payable
- where a percentage of sales calculation is adopted, as opposed to a fixed fee per item sold, the contract must define how the net sales value shall be calculated
- the rates at which royalties are payable
- any limitation on period over which royalties are payable
- where a minimum royalty is agreed this is to be set out clearly in the contract
- details of manner and timing of payment, currency and/or currency conversion
- permitted and non-permitted deductions, tax withholding and so on
- obligation on the part of the knowledge exploiter to keep proper records and make them available when requested
- obligation on the knowledge exploiter to put some degree of effort into commercial exploitation.

The amount of royalty will be based on the relative bargaining strengths of the parties and their perception of the value of any pre-existing rights to the value of the final commercialized product. Setting off of the costs of the research contract will usually be appropriate in these situations, as may be royalty reduction in the event that patent protection is not available in particular sales territories.

APPENDIX 1

The Outsourcing R&D Toolkit

P. A. Sammons, *The Outsourcing R&D Toolkit*, Gower, 2000, ISBN 0 566 083140.

CONTENTS

- RDT11 Standard R&D contract letter
- RDT12 Unsuccessful tender – decline letter
- RDT13 Project memo
- RDT14 Variation to contract – letter
- RDT15 Confirmation of discharge of contract – letter
- RDT16 Research vendor evaluation checklist (use by visiting group)
- RDT17 Confidentiality agreement – third party sends information to us
- RDT18 Confidentiality agreement – third party hands information to us at meeting
- RDT19 Confidentiality agreement – we send information to third party
- RDT20 Confidentiality agreement – we hand information to third party at meeting
- RDT21 Request for reference – letter and questionnaire
- RDT22 Confirmation of satisfactory completion of assessment – letter
- RDT23 Contract close-out checklist

13. Appendix: Instructions for use of template documents

APPENDIX 2

Project Memo

Project Memo

Commercial in confidence

KNOWLEDGE PROJECT TITLE

To: .... Project memo # ....

Via: .... Page 1 of..

From: .... Reply requested? YES NO

Date: .... Date reply required ....

Subject: ....

Fax no:/E-mail to: ....

This message is not a Contract Variation/change authorization. Contract changes (changes to Contract specification, scope, programme, prices etc.) must be made in accordance with the terms of the Contract.

1. REFERENCES

2. DISCUSSION

3. ACTION REQUIRED

4. ATTACHMENTS

APPENDIX 3

Watch Your Service Bills!

In December 2003 PriceWaterhouseCoopers settled charges that it had over-billed clients in the US amounting to hundreds of millions of dollars by not passing on rebates received from travel agencies. At the time of writing this book, class action suits related to invoicing were still pending against KPMG, Ernst & Young and legal firm Ogilvy & Mather. Companies and other organizations spend enormous sums on services. It is reckoned that the average company today spends 33% of total external expenditure on services, which encompasses process outsourcing, consultancy advice and legal services. But this figure can vary widely. For financial services firms it can be as high as 80% of external expenditure. Services spending continues to rise at an annual rate of 3.5%, according to CAPS research, a firm that monitors corporate purchasing.

For client organizations there are significant and relatively easy savings to be made by better monitoring and controlling service bills. Whilst outright fraud is quite rare, inadvertent overcharging is alarmingly common. There are four principle reasons:

- simple carelessness
- inadequate cost tracking systems within service-supplier organizations
- failure to understand the agreement
- client too busy to properly scrutinize bills.

Errors, whether intentional or not, can cost individual clients quite staggering sums. Stephen Broderick, chief operating officer at a London-based company that audits advertising-agency compliance with client contracts, cites a 2003 audit where a US advertising agent mistakenly invoiced a client $84 000 for 12 hours of secretarial work. In another example, cited by Judie Bronsther, president of US firm Accountability Services, a law firm's partners were found to be charging all their meals and transport costs to a client, regardless of the time actually spent on the work!

Companies in particular can fall foul of the scope-creep problem identified in Chapter 8. Some professional services firms want to extract as much follow-on business as they can from their clients, sometimes because these clients are a soft touch, and sometimes because as professionals they feel it appropriate to try to assist in ever greater ways. So, management consultants for example, having had privileged access to client organizations' senior managers and strategic plans, as well as their operating practices and plans, are well placed to identify new problems which, of course, they can then assist in resolving! Bear in mind that whilst a management consultant works in your organization, they may simultaneously be working on 'proposals' related to other opportunities they have identified. If they can present these opportunities as urgent and quickly present a proposed solution to the client, the client is then under considerable pressure to react (note the word react!) and probably will not test the market, out of a sense of gratitude to the consultant who has taken the time and trouble to identify other 'problems' as well as the perceived need to speedily deal with the 'urgent' problem.

One solution to this perennial difficulty is to draft and enforce a tight contract with the professional services supplier. Keep them focused on the problem in hand, and be cautious about the freedom they are given to wander around the client organization. Over-billing can be prevented in the first place by developing a detailed billing policy and spelling this out in the contract. Fixed, milestone-based payments are one method to control the problem. The author, when faced with this problem, devised a simple and effective approach to billing, by insisting that all invoices should routinely include the following:

TIME AND MATERIALS-TYPE WORK

1. Project/contract title
2. Contract number (where available)
3. Name of consultant(s) used
4. Grade of consultant(s) used
5. 'Rate card' rate – if applicable – per consultant
6. Hours worked – plain time – overtime (and overtime rates, if different)
7. Date services rendered – if split over different fee periods
8. Discount applied – where relevant
9. Expenses (if relevant) and receipts attached
10. Net total
11. VAT (or other relevant taxes)
12. Gross total

13. Precise payment address
14. Name of client organization controlling manager.

FIXED STAGE PAYMENTS-TYPE WORK

As above, 1, 2, and 8–13. In addition

1. Stage/milestone delivered
2. Outline of deliverables supplied to client organization.

Where necessary such information can be included in attachments or continuation sheets, providing they are clearly referenced on the front page of the invoice. The contract clause covering invoicing should include precise details about the format of bills as suggested above.

Putting the heat on lawyers!

Companies are more willing today to challenge legal bills. Traditionally they paid bills without question because they thought they could not afford to harm the lawyer–client relationship. Today, managers recognize that legal fees should be managed as closely as any other cost centre. It is usually a mistake to expend effort haggling over fee rates, rates that could be the least important component of the ultimate cost build. Instead client organizations should look at other factors, such as how the legal firm staffs a project, how it manages the overall assignment (brief) and how the firm bills for expenses. Items to consider in a billing policy for legal eagles are suggested below:

- The firm should seek approval for research work that requires five or more billable hours.
- Do not pay for more than one lawyer to attend a routine hearing.
- Define what expenses you are willing to pay for. Clients should not pay for secretarial support or fax costs.
- Do not pay for the firm's overhead – heating, business rates or rent, for example. (Admittedly there may be a hidden – or even a disclosed – overhead rate that the client will pay, but the idea is to secure as much comparability between fee rates as possible, and avoid comparing some rates, that are all-inclusive, with others that exclude certain elements that will later be claimed separately.)
- Require the firm to maintain *continuity of staffing* for a project. A client should not have to pay the additional costs associated with bringing a new lawyer up to speed on a project.

- Require the firm to notify you in writing in advance of any increase in billing rates.
- Billing 'increment triggers' should be small: about every six minutes. Without this level of 'granularity' in billing, a client may find it is billed for 15 minutes of partner time for a two-minute phone call.
- Require 'most favoured nation' status, where you pay the lowest hourly rates charged to other clients of a similar size.

Hire service experts

This book has been about *buying knowledge*, an area acknowledged as being difficult to specify, and requiring a suitable amount of up-front definition and planning. Although this is not specifically a book about procurement, a number of the strategies and procedures recommended in the book will often in practice be carried out by the organization's procurement operation. But it is a basic philosophy of this book that organizations should apply the same rigour to buying services as to physical goods. It is hard to put a value on advice. It is hard to put a value on knowledge. We cannot simply specify a quantity and then shop around for a price. Because quality can vary widely from provider to provider, unit cost (such as hourly fee rates) is relatively unimportant.

It is also a fact that service contracts are often very complex. A project may require many component parts and intricate pricing, often including one or all of the following:

- milestone payments
- time and materials pricing
- financial reward mechanisms
- contingency costs.

These things can be difficult to capture in an e-procurement system, which means that high-value services often bypass the procurement process altogether.

There is also a question around how client organizations account for service costs. Money spent on materials used in manufacturing affects a businesses cost of sales directly. Saving 10% on these costs yields an obvious and immediate benefit to the bottom line. Services, by contrast, are often budgetary. If there is a £10 000 budget for brand and marketing, a marketing manager will probably spend the full amount, if not because of a direct need then because of fear of losing the budget allocation in the next financial year.

A useful counter to these problems is for the client organization's procurement department to hire its own services experts. For example a procurement professional who has specific legal services buying experience or a former consultant who knows how the market operates – even a former lawyer who knows in detail how much certain types of legal procedure should cost. By working closely with internal users of services, the procurement group should be able to identify key service areas and key external service suppliers, and then focus negotiations on these firms. To give the client organization an advantage in negotiations, a good procurement team will, certainly for a high-value project, prepare for negotiations by modelling the cost of the service for a supplier. The procurement negotiator will then present its assessment of what the supplier's costs should be, add in a reasonable profit and then tailor negotiations closely to this proposition.

Just as important as managing the supplier is keeping the client organization's own managers in check. A formalized *demand challenge* process is an extremely valuable way of reducing costs of proposed external projects. Refer again to Figure 8 on page 65. This formalized process should take place as close to inception of the proposed project as is possible. Test every assumption made. Always assume that ultimately the project may be unnecessary, or that the scope can be reduced.

Index